Fossil Fuel Hydrogen

William J. Nuttall · Adetokunboh T. Bakenne

Fossil Fuel Hydrogen

Technical, Economic and Environmental
Potential

 Springer

William J. Nuttall
School of Engineering and Innovation
The Open University
Milton Keynes, UK

Adetokunboh T. Bakenne
School of Engineering and Innovation
The Open University
Milton Keynes, UK

ISBN 978-3-030-30907-7   ISBN 978-3-030-30908-4   (eBook)
https://doi.org/10.1007/978-3-030-30908-4

This Springer imprint is published by the registered company Springer Nature Switzerland AG
The registered company address is: Gewerbestrasse 11, 6330 Cham, Switzerland

To Maggie and Lola

Notes and Acknowledgements

We are most grateful to numerous people who in various ways greatly helped our understanding of these issues. In particular, we would like to record our thanks to P. V. Aravind, Nick Butler, Bernardo Castro-Dominguez, Richard Clarke, Arnab Chatterjee, Chi Kong Chyong, Steven Feldman, Charles Forsberg, Bartek Glowacki, Damien Hawke, Roger Hunter, Mohammed Kalbassi, Nikolaos Kazantzis, Satheesh Krishnamurthy, Michael Lewis, Alaric Marsden, Nazim Muradov, Pierre Noel, Jaap Oldenziel, David Reiner, Dave Robson, Dan Sadler, Simon Schaeffer, Herie Soto, Mike Timko, Matthew Tipper and Margot Weijnan.

We are especially grateful to those that provided assistance in various ways directly associated with specific chapters. Their contributions are discussed in the respective chapters.

Those that have offered assistance or inspiration do not necessarily agree with the findings and opinions presented. Any errors or omissions are the responsibility of the authors alone.

Note Concerning Images

Efforts have been made to establish and contact copyright holders for all images presented in this book. We are most grateful to all the various rights holders who have kindly granted permission for reproduction. Despite our best endeavours, there may be instances where the rights of third parties have been overlooked. In such cases, we apologise and we ask that rights holders make contact and we will endeavour to resolve matters.

Contents

Acronyms

ACO	Automobile Club de l'Ouest
BCM	Billion cubic metres
BEV	Battery electric vehicle
BP	British Petroleum
CAGR	Compound annual growth rate
CCGT	Combined cycle gas turbine
CCS	Carbon capture and storage
CCUS	Carbon capture, Utilisation and Storage
CMR	Catalytic membrane reactor
CO	Carbon monoxide
CU	Carbon Utilisation
DAC	Direct air capture
DMFC	Direct methanol fuel cell
DOE	Department of Energy (USA)
ECBMR	Enhanced Coal Bed Methane Recovery
ENS	Enhanced natural sink
EOR	Enhanced Oil Recovery
EU-ETS	European Union—Emissions Trading Scheme
EV	Electric vehicle
FC	Fuel cell
FCEV	Fuel cell electric vehicle
GHG	Greenhouse gas
H2	Hydrogen
HDS	Hydrodesulfurisation
HEV	Hybrid electric vehicle
HHV	Higher heating value
HP	High pressure
HTS	High-temperature superconductor
HYCO	Hydrogen–carbon monoxide
ICE	Internal combustion engine

IEA	International Energy Agency
IOC	International oil company
IT	Information technology
LH2	Liquid hydrogen
LHV	Lower heating value (equivalent to net calorific value)
LNG	Liquefied natural gas
LPG	Liquefied petroleum gas (also "autogas" or "autopropane")
MRI	Magnetic resonance imaging
NG	Natural gas
NOC	National oil company
O&M	Operations and Maintenance
ODS	Oxidative desulfurisation
OECD	Organisation for Economic Co-operation and Development
OPEC	Organization of Petroleum Exporting Countries
PAFC	Phosphoric acid fuel cell
PEM	Proton exchange membrane
PHEV	Plug-in hybrid electric vehicle
POX	Partial oxidation
PPMW	Parts per million by weight
PSA	Pressure swing adsorption
SCW	Supercritical water
SCWU	Supercritical water upgrading
SER	Sorption enhanced reforming
SMR	Steam methane reforming
SOEC	Solid oxide electrolysis cell
SOFC	Solid oxide fuel cell
STP	Standard temperature and pressure
TGA	Thioglycolic acid
WGS	Water gas shift

Chapter 1
Introduction—The Hydrogen Economy Today

This book is focussed on the future. How can humanity ensure prosperity and mobility in the decades to come without irreversibly damaging our planet? One key imperative will be to reduce drastically the emission of harmful greenhouse gases, and most especially carbon dioxide. Today's mobility, based upon the combustion of petroleum, is a key component of concern going forward. Another climate challenge comes from the use of natural gas in domestic heating. Many voices argue that the future lies in electrification, the logic being that ways are known to generate electricity with very low harmful emissions, such as via renewable sources including wind and solar. Furthermore, the growing numbers of battery electric vehicles can allow one to imagine that the end of oil is in sight. Such a future may indeed occur, but we suggest that the end of fossil fuels is not inevitable and perhaps not even desirable if the risks to the climate can be avoided. The electrification path is not necessarily the only path associated with a low-carbon future and in this book we explore another— one that makes use of hydrogen as a future energy carrier and that seeks to minimise greenhouse gas emission via carbon capture, utilisation and storage. We use the term energy carrier, as opposed to "fuel" to make clear that hydrogen must be produced, using some other energy resource, as molecular hydrogen does not exist in sufficient accessible abundance on Earth.

Much attention has been given to the possibility of producing hydrogen from renewable energy sources, but in this book we deliberately give emphasis to an alternative: the continued production of hydrogen from fossil fuels, such as natural gas, but in ways that can be developed so as to minimise greenhouse gas emissions. Such a path of investigation will lead us to assess the merits of a widely held perception, especially prevalent among academic hydrogen economy researchers, that fossil-fuel-based hydrogen production methods are inevitably "*low tech, polluting and without significant potential for innovation*". In this book, we shed light on the realities of such methods in hydrogen production; we assess their future prospects and, where appropriate, we challenge false perceptions.

Figures from Ref. [1] reprinted under licence (number 4338730347221) from the International Journal of Hydrogen Energy.

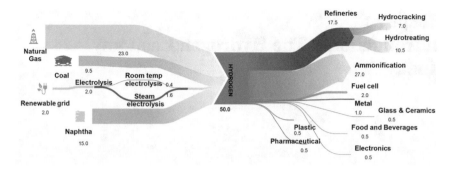

Fig. 1.1 Average supply and demand of global hydrogen supply, in million metric tons. Data assembled from multiple sources (2004–2013) [1]. The oil refining terms "hydrocracking" and "hydrotreating" are defined in the text

In this book, we take a whole systems approach and we consider current options and scenarios for the development of the hydrogen industry. We consider various strategic choices faced by both hydrogen producers and consumers. In so doing, we hope to reveal useful opportunities for the development of a robust well-functioning and growing hydrogen economy consistent with minimising harmful environmental impacts. Additionally, we seek to inform policy-makers on future trends for hydrogen commercialisation especially those emerging from today's industrial reality.

While the initial motivations for renewable energy came from early 1970s concerns surrounding oil supply security, more recently the driving motivation has been a desire to decrease the greenhouse gas emissions associated with transport and mobility. In this book, we shall describe renewables-based approaches to hydrogen production as the "Green Hydrogen" paradigm. Today "Green Hydrogen" represents a vision usually associated with renewable electricity generation, hydrogen production by electrolysis, new hydrogen supply chains, on-vehicle hydrogen storage and advanced fuel-cell-based electromechanical power trains for road vehicles.

Hydrogen is a well-established product of the industrial gas industry, and while its scale currently falls short of that associated with some scenarios for the much-vaunted "hydrogen economy", it is already a significant and important industrial activity. In this book, we shall sometimes refer to these well-established industrial activities as "Mature Hydrogen". In so doing, we avoid the terms "Brown Hydrogen" and "Blue Hydrogen" that are sometimes used so as to contrast with renewables-based "Green Hydrogen". We find the terms "Brown", for hydrogen from coal, and "Blue" for hydrogen from natural gas to be rather too simplistic, and perhaps even pejorative, for our purposes. The vast majority of today's hydrogen is sourced via Mature Hydrogen processes (see Fig. 1.1). Of this, a large fraction is associated with transport and mobility, as it is consumed by the petrochemical industry for removing sulphur from sour crude oil, and for producing less viscous petroleum-based vehicle fuels; this will be discussed further later in the chapter. The other major use for Mature Hydrogen is in fertiliser (ammonia) production.

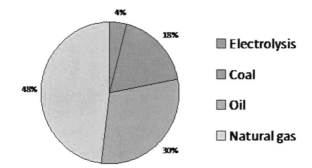

Fig. 1.2 Global hydrogen production [2, 3]. Note the electrolysis segment is 4% and the sequence in the key and diagram runs clockwise from there

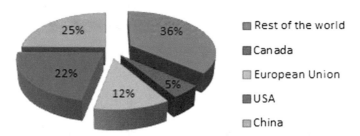

Fig. 1.3 Annual global hydrogen production (total is approximately 50 million metric tons). From: Bakenne and Nuttall, primary sources described therein [1]. Note the Rest of the World segment is the largest (36%), and the sequence in the key and diagram runs clockwise from there

Figure 1.1 reveals the scale of the Mature Hydrogen production industry today and further illustrates how this industry swamps the "Green Hydrogen" (renewable electricity to fuel cell) value chain. Such green flows represent only a tiny proportion of total hydrogen (by mass). In Fig. 1.1, the proportion from renewable electrolysis is shown as being 2%, and Fig. 1.2 shows the total coming from all electrolysis (using renewable and non-renewable electricity) to be 4%. Clearly, the proportion of hydrogen coming as Green Hydrogen from electrolysis using renewable electricity is currently very small.

As things stand today, Mature Hydrogen dominates hydrogen production, and hence, any process improvement within that industry, such as measures aiming to reduce greenhouse gas emissions, will have far more absolute beneficial impact than an equivalent proportionate improvement in the contribution from renewables-based Green Hydrogen, i.e. any 1% incremental improvement in the Mature Hydrogen sector would have an impact, in the short-term at least, equivalent to a 25% improvement to Green Hydrogen methods. These realities will persist for some time to come even in scenarios of significant growth in Green Hydrogen production. As such, Mature Hydrogen will clearly be dominant in all short-to-medium-term hydrogen futures (Fig. 1.3). This near-term reality in part motivates this book, but the question

Fig. 1.4 London Hydrogen-fuelled fuel cell electric bus 2015. *Source* Author (WJN)

Table 1.1 Greenhouse gas emissions: hydrogen versus petroleum assuming no carbon capture [4]

Power units	Type of fuel	GHG emission (g/kWh)
Fuel cell (for forklifts)	Hydrogen from natural gas	800
	Hydrogen from wind	200
IC engine	Gasoline	1250
	Diesel	1300

then becomes might the fossil fuel feedstock be used, over the longer term, in more environmentally responsible ways?

Across the world, there is much research interest in the possibility of a disruptive innovation in which hydrogen might be generated using intermittently surplus renewable energy sources. This is the renewables-based Green Hydrogen paradigm. However, far less technology policy research is devoted to examining incremental innovation in established mature methods for hydrogen production. The dominant established, and mature process is steam methane reforming (SMR), as led by industrial gas companies.

Table 1.1 highlights the greenhouse gas emissions' impacts of Mature Hydrogen and Green Hydrogen and compares these with the emissions from conventional petroleum-fuelled internal combustion (IC) engines.

Table 1.1 makes clear that even today, Mature (natural gas originated) Hydrogen is a lower greenhouse gas emissions mobility option than petroleum fuels. Clearly, wind energy generated hydrogen (as an example of Green Hydrogen) today scores better than Mature Hydrogen, However, we shall go on to consider the opportunities by which Mature Hydrogen might reduce its greenhouse gas emissions via incremental innovation building on the strong industrial base introduced earlier and elaborated on further in Chap. 3.

To be clear, today's unabated methods of hydrogen production from natural gas and other fossil fuels represent significant sources of harmful greenhouse gas emissions. Indeed, the US Department of Energy has observed that roughly 5% of all US transport emissions relate to the use of hydrogen in vehicle fuel processing (hydrocracking and hydrotreating—defined later in this chapter) [5]. Hence, even in the absence of any transition from petroleum to hydrogen as a vehicle fuel, cleaning up the production of hydrogen could have a significant effect on global GHG emissions.

1.1 Perceptions and Reality

Arguably, hydrogen is much misunderstood. Indeed, it seems likely that the industrial gases and international oil companies could together be a major part of the solution to looming global problems rather than, as presently, being widely perceived to be solely part of the problem.

Hence, in this book, we intend to explore the idea that Mature Hydrogen production is already a "material" (meaning substantial and worthy of significant attention) business open to further innovation, and capable of future high-impact contributions to global policy goals associated with more environmentally responsible behaviours and economic growth.

1.2 The Uses of Hydrogen Today

Approximately 50 million metric tons of hydrogen are produced globally each year [1, 4, 6] As illustrated in Fig. 1.5, the hydrogen is used for petroleum refining, fertiliser production and methanol production. In addition, hydrogen is used by the semiconductor industry, in window glass manufacture, in metallurgy and for food hydrogenation purposes. The plans for hydrogen use in fuel cell applications are widely publicised and celebrated (see Fig. 1.4), but for now this remains a very small part of the hydrogen story.

Despite the substantial scale of the global hydrogen industry, the full scale commercialisation of hydrogen as an energy carrier has not been achieved despite the introduction of the phrase "hydrogen economy" more than 40 years ago. It is expected hydrogen will play a key role especially in the decarbonisation of the transport sector and elimination of the tailpipe emissions from vehicles [8], but as we have seen, so far this component is small in comparison to the established hydrogen economy.

Much of today's industrial experience in hydrogen production relates to its role in the manufacture of petroleum fuels. In all scenarios, such fuels will continue to play a major role in the global energy mix for many years to come. Of course, the unabated use of fossil fuel is associated with the looming catastrophe of serious

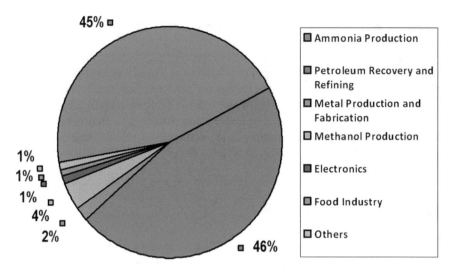

Fig. 1.5 Global hydrogen use. Ammonia production is 45%, and other entries follow the key in a clockwise manner [7]

environmental harm. The environmental concerns span across exploration, refining and most especially end use of such fuels. The environmental issues run wider than climate change—one cause of harm associated with petroleum use has been the historical production of sulphur dioxide in gasoline combustion.

Sulphur in fuel degrades vehicle emission control systems; it damages human health, and it also causes acid rain with the potential to poison lakes, rivers, forests and crops [9, 10]. In order to address these concerns, many countries have imposed strict regulations to minimise the sulphur content of fuels. Crude oil with a high sulphur content is said to be "sour". The desulphurisation of petroleum and petroleum fractions is almost universally accomplished by the catalytic reaction of hydrogen with sulphur compounds in the charge stock to produce hydrogen sulphide (H_2S). This process forms a major part of what is often known as "hydrotreating". The H_2S produced by the process is then readily separable from the oil being processed [11]. Such hydrodesulphurisation operations are in widespread use in the petroleum refining industry. The technology of hydrodesulphurisation is now well established, and petroleum stocks of every conceivable molecular weight range are now being treated to remove sulphur. A key driver for growth in the hydrogen demand comes from the petroleum refining industry as it seeks to meet the requirements of increasingly stringent legislation concerning the maximum sulphur content in fuels. A second major driver of hydrogen demand concerns the shift in recent decades to lower-quality heavier crude oils in the upstream petroleum industry. These crude oils require hydrogen for "hydrocracking" of the oil to lighter molecules before downstream use. Furthermore, all this comes against a background of increasing oil consumption in developing economies (i.e. China, India).

So when considering the possibility of hydrogen for use in low carbon mobility, it should be remembered that today, hydrogen plays an essential and growing role in producing improved petroleum-based transport fuels (as discussed further in Chap. 2 and subsequently). Whether this strong existing linkage between the hydrogen industry and mobility might be evolved into something more compatible with a low-carbon future remains to be seen. Such thinking again forms part of the motivation for this book.

1.3   Hydrogen Demand Growth

The global hydrogen generation market continues to grow driven by increasing demand. It is currently over $100 billion, and it is estimated that it will reach $200 billion by 2025 [12]. In 2014, 88% of global hydrogen production was related to the needs of the petrochemical industries. The other 12% of the global hydrogen produced having been taken by merchant consumers [4]. Merchant actors in the hydrogen sector can make, or buy, hydrogen and they sell hydrogen. Typically, they do not themselves use hydrogen for industrial processes. That is left to their customers. If such a customer prefers not to deal with merchant providers for any reason, they can source their own hydrogen. Such production is said to be "captive". Hydrogen consumption growth in the period up to 2018 has been indicated to be 5–7% [13, 14]. As shown in Fig. 1.6, the captive production market is expected to rise from $90.81 billion in 2014 to $118.11 billion by 2019, with a compound annual growth rate (CAGR) of 5.4% [4]. The Asian market is expected to have sourced more hydrogen from captive production in the period 2014–2017 [5] due to demand increase by oil refineries. It is also noteworthy that Asian countries, such as China, India and Sri

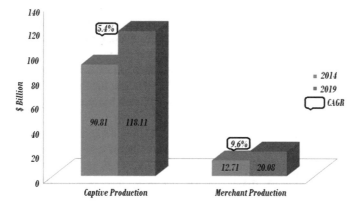

Fig. 1.6 Hydrogen captive production versus merchant production, 2014 as measured and 2019 as predicted ($ billion) adapted from [4]. CAGR refers to compound annual growth rate. Left-hand columns correspond to 2014

Lanka, have proposed to tighten their sulphur standards for vehicle fuels. This will further drive up hydrogen demand. The market share of merchant production market has also been expected to rise from $12.71 billion in 2014 to $20.08 billion in 2019, corresponding to a CAGR of 9.6% [4]. Globally, this industry is dominated by a set of competing industrial companies comprising: Air Products and Chemicals Inc. (USA), Air Liquide (France), Linde AG (Germany) incorporating BOC Ltd (UK), and Praxair (USA). Globally, there is good competition between these companies in generating and distributing hydrogen to individual customers. In some specific territories, however, competition is more limited and some level of market power can be expected. That said, the market is generally contestable and any hydrogen user unable to purchase from merchant providers could always embark on developing their own captive capability.

In 2018, Praxair and Linde joined forces in what has been described as a "merger of equals" creating a company with a market capitalisation of around US$ 90 billion [15]. The new combined company usurps Air Liquide's former status as the world's largest industrial gas company.

1.4   Fossil Fuel Hydrogen: Low Tech?

The fossil-fuel-based hydrogen industry is large-scale and highly technical. It involves advanced chemical and mechanical engineering knowledge. Fundamentally, it is an engineer's world. In contrast, the world of green renewables-based hydrogen has included a wide range of types of individual ranging from concerned scientists and engineers to social scientists and campaigners. While generally diversity of thinking leads to improved decision-making and improved outcomes, when the issue is highly technical in nature, engineering-oriented teams can make rapid progress, providing the goals are clear and widely supported.

The technologies of hydrogen production from natural gas, such as steam methane reforming (SMR) involve important proprietary knowledge and valuable intellectual property (e.g. in engineering design and chemical catalysts). In particular, there is a track record of innovation such as the modification of nozzles in SMR burners to reduce nitrous oxide emissions… in summary: fossil fuel hydrogen production is "high tech". We shall consider the role of innovation in Mature Hydrogen production further in Chap. 3.

At the risk of overgeneralisation and simplification, it might be helpful to note that our previous work has indicated that the production of hydrogen from fossil fuels is currently 3.7 times cheaper than equivalent hydrogen produced via electrolysis [1]. The production of hydrogen via electrolysis is a key component of proposals to manufacture hydrogen using otherwise surplus renewable electricity. Of course the statement that fossil-fuel-based hydrogen is cheaper relies on an assessment of the cost of the electricity used. Some might say that the renewable electricity in question is extremely cheap. They will be thinking in marginal cost terms or in terms of prices available in territories that have strongly supported renewables deployments.

We would take the view that such cost assessments are not appropriate for good decision-making and a more holistic appreciation of costs is needed, as fixed and sunk costs must be included. Alternatively one can take an entirely different approach and use actual electricity prices for the market as a whole in the hope that can in some way capture total costs. This is what we did in our previous work [1]. If one were to approximately equalise the environmental impacts of Mature Hydrogen and Green Hydrogen by, for instance, the use of carbon capture and storage (CCS) then this cost ratio drops to 1.8. That is, for roughly equal environmental credentials Green Hydrogen, production is still nearly twice the price of Mature Hydrogen—even in the narrow framing of the studies cited by our previous work [1]. That initial observation has led us to be interested in the evolution to a low-carbon hydrogen economy from its current industrial base serving the changing needs of the oil and chemical sectors.

1.5 Fuels in Transition

Even among communities most open to the notion that natural gas could provide the basis for a low-carbon hydrogen-based future for mobility and heating, there is a much repeated view that this is just a temporary option, and merely a transition to something more sustainable and enduring. While at a certain level, this must be true, as after all, nothing lasts forever; at another level, we suggest that it is unhelpful to good policy-making and strategy development.

When the history of these things is written, it seems likely that the age of oil will have lasted for 150 years from the late nineteenth to the early twenty-first century. In retrospect it will appear to have been broadly coincident with the twentieth century. We doubt very much, however, if anyone will look back on oil as having been merely a "transition fuel". Natural gas is an abundant resource made more important by innovations in extraction methods including, but not limited to, hydraulic fracturing or "fracking". We can posit a low carbon hydrogen age, across much of mobility and heating that might also endure for more than 100 years and which could be based on natural gas. It could enable a low-carbon and secure future at lower cost and with far less effort and disruption than might be expected to be seen in renewables-based electrification scenarios.

One can look at the future of mobility and see a way ahead inspired by the rapid progress of renewable electricity capacity expansion as pushed successfully by environmentally minded energy policy. One can also, however, look at the remarkable transformation of the global natural gas industry from a pipeline constrained commodity in an ever-decreasing number of global regions twenty years ago to today's current realities as a globalising and ever more high tech moderately low-carbon energy solution sourced via new means, such as fracking, from a broadening range of territories. Both are impressive achievements, but not without controversy. The consequential question considered by this book is which is going to have the most significant impact on the emerging hydrogen economy (Fig. 1.7).

Fig. 1.7 Russian ice-capable liquefied natural gas tanker the Christophe de Margerie (right) docked in Sabetta, Russia, 30 March 2017. As climate change melts the northern ice, a new safe sea route emerges. *Source* Friedemann Kohler/dpa/Alamy Live News

We suggest that the future of natural gas should not be seen as merely a transition towards future more sustainable options, rather we suggest that it might be seen as beneficial solution that can endure for the rest of our lives and beyond and as a route to a secure and responsible future accessible to all, including to countries with very limited wealth. Of course, others might point to the falling costs of renewables and suggest that renewable-based electricity could be an affordable proposition for all. One day that renewables-dominated future might come, but the current cost data, when considered holistically, reveal that we currently very far from such goals. We here posit that, while natural-gas-based Mature Hydrogen might indeed be a transition fuel, it also has the potential to be much more enduring than that.

1.6 Future Prospects for Fossil-Fuel-Based Hydrogen Production

Although today's Mature Hydrogen production is rightly regarded as contributing to the problem of global climate change, interestingly it need not necessarily be such an environmentally harmful process.

Table 1.2   Relative attributes of hydrogen production methods—authors' assessment

Fossil-fuel-based Mature Hydrogen	Green renewables-based hydrogen
Industry thought leadership	Government/academic thought leadership
Technology/market pull	Policy push
Potential for evolutionary emergence from today's mobility economy	Potential to disrupt today's mobility economy
Much at high-technology readiness levels	Much at low-technology readiness levels but exhibiting rapid progress
Perceived to be greenhouse gas emissions problematic	Perceived to be greenhouse gas emissions free

Looking ahead, it seems probable that the long-term potential of hydrogen as a low-CO_2 energy carrier will not, by itself, be sufficient to prompt industrial innovations that are not otherwise economically sustainable. We note, in addition, the difficulties encountered worldwide in attempts to establish a meaningful price for greenhouse gas emissions. Such a price might have been expected to spur innovation, but it has not yet done so. Rather than look only to disruptive innovation, we suggest that a progressive hydrogen economy might emerge more incrementally from the global hydrogen industry as it already exists today. This represents an alternative way ahead to those deriving from technology push either in the form of policy aspiration, or as it motivates university-led research and development. As such, Mature Hydrogen and Green Hydrogen differ in key respects (Table 1.2).

1.7   The Hydrogen Council

The Hydrogen Council, launched at the World Economic Forum in Davos, Switzerland in 2017, describes itself as a global initiative of leading energy, transport and industry companies with a united vision and long-term ambition for hydrogen to foster the energy transition [16]. In November 2017, the council published a roadmap report entitled *Hydrogen Scaling Up* [17].

The *Scaling Up* report represents a powerful overview of the potential for hydrogen to contribute to a low carbon future. Many of the messages are consistent with the ideas stressed in this book, but there are some minor, and one major, points of difference. One of the more minor matters is that the authors of this book see a clear role for nuclear energy to contribute to a low carbon future. The Hydrogen Council document barely mentions nuclear energy and when it is mentioned nuclear power is lumped together with fossil fuel generated electricity. Given the fundamentally different economics and carbon credentials of nuclear power and fossil fuel generated electricity such an aggregation would appear to be a somewhat odd choice. Indeed, high capital costs, the apparent need for public subsidy, good supply security and low-carbon credentials would suggest that nuclear power should be considered alongside renewable power, but that is not the Hydrogen Council's choice.

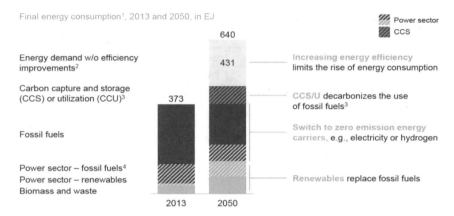

Fig. 1.8 Hydrogen Council overview of the possible growth of global energy consumption to 2050. *Source* Hydrogen Council (hydrogencouncil.com) [17]

The substantial point of difference between the Hydrogen Council's *Scaling Up* report and this book, however, lies in the expected role to be played by renewables in future hydrogen production. The Hydrogen Council goes so far as to say that "*the energy supply needs to transition to renewable sources* [17]", and the Council observes that hydrogen enables large-scale renewables energy integration and power generation [17]. While such statements might turn out to be true, in this book we choose to emphasise a more evolutionary approach by which the international oil companies, having become international natural gas companies, now become international hydrogen companies—producing and selling hydrogen primarily obtained from natural gas, but generated in ways that have minimal environmental impact. We see this being an easier, cheaper and similarly clean way ahead.

Despite our points of divergence from the opinions of the Hydrogen Council, we commend them for having produced a most informative and interesting policy document. We close this chapter with reference to two figures in the Council's *Scaling Up* report. As one can see from Fig. 1.8, the Hydrogen Council does see a clear and substantial role for fossil fuel hydrogen in conjunction with CCUS (denoted CCS/U in the figure) despite the Council's enthusiasm for a shift to renewables-based Green Hydrogen.

Within the energy growth posited in Fig. 1.8, the Hydrogen Council sees scope for an order of magnitude increase in hydrogen demand out to 2050 (Fig. 1.9).

Whatever the merits of fossil-fuel-based "Mature" or renewables-based "Green" hydrogen production, it is clear that the world has before it an opportunity for an important change in the energy system. Whether the world chooses to make such a shift will involve consideration of issues of economics, technology and politics. In large part, such decisions will be shaped by the growing realisation that as regards

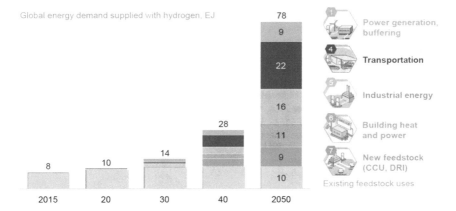

Fig. 1.9 Hydrogen Council proposed increase in global hydrogen demand to 2050. *Source* Hydrogen Council (hydrogencouncil.com) [17]

global climate stability something substantial must be done, most especially in areas of heating and mobility. It is in those specific areas that hydrogen has a particularly compelling role to play, as the rest of this book will consider.

References

1. Bakenne, A., W. Nuttall, and N. Kazantzis. 2016. Sankey-diagram-based insights into the hydrogen economy of today. *International Journal of Hydrogen Energy* 41 (19): 7744–7753.
2. Lemus, R.G., and J.M. Martínez Duart. 2010. Updated hydrogen production costs and parities for conventional and renewable technologies. *International Journal of Hydrogen Energy* 35 (9): 3929–3936.
3. Power Generation 2018. 9th September 2018. Available from: http://www.airproducts.com/Industries/Energy/Power/Power-Generation/faqs.aspx.
4. MarketsandMarkets. *Hydrogen generation market by geography, by mode of generation & delivery, applications and technology—Global trends & forecasts to 2019 in Hydrogen Generation Market 2014*, 1–156. Washington: MarketsandMarkets Inc.
5. Muradov, N. 2017. Low to near-zero CO$_2$ production of hydrogen from fossil fuels: Status and perspectives. *International Journal of Hydrogen Energy* 42 (20): 14058–14088.
6. Hydrogen—The key energy enabler, in *Gasworld*. 2013, 38–40.
7. Thompson L., B.F., Burns L, Friedland R, Kiczek E, Nozik A, Richmond G, Shaw R, Wilson D. 2013. *Report of the Hydrogen Production Expert Panel to HTAC* 2013. Washington DC: United States Department of Energy.
8. Pasion, C., M. Amar, and M. Delaney. 2016. *Inventory of New York City Greenhouse Gas Emissions November 2014*, in *Mayor's Office of Long-Term Planning and Sustainability*. New York.
9. Kida, Y., et al. 2014. Combining experiment and theory to elucidate the role of supercritical water in sulfide decomposition. *Physical Chemistry Chemical Physics* 16 (20): 9220–9228.
10. Gonzalez, L.A., et al. 2012. oxidative desulfurization of middle-distillate fuels using activated carbon and power ultrasound. *Energy & Fuels* 26 (8): 5164–5176.
11. Schuman, S.C., and H. Shalit. 1971. Hydrodesulfurization. *Catalysis Reviews* 4 (1): 245–318.

12. Global Market Study on Hydrogen. 9th September 2018. Available from: https://www.persistencemarketresearch.com/market-research/hydrogen-market.asp.
13. Transport and the Hydrogen Economy. 2014. Available from: http://www.world-nuclear.org/info/Non-Power-Nuclear-Applications/Transport/Transport-and-the-Hydrogen-Economy/.
14. Hydrogen. 2015. *Chemical economics handbook 2015*. Available from: https://www.ihs.com/products/hydrogen-chemical-economics-handbook.html.
15. Cockerill, R. 2018. *A new leader crowned: Praxair and Linde to complete $90bn merger of equals.* Gasworld, 22 October 2018. Available from: https://www.gasworld.com/praxair-linde-to-complete-90bn-merger/2015680.article.
16. Hydrogen Council. 2018. Available from: http://hydrogencouncil.com.
17. *Hydrogen scaling up-a sustainable pathway for the global energy transition.* 2017, Hydrogen Council. Available from http://hydrogencouncil.com/wp-content/uploads/2017/11/Hydrogen-scaling-up-Hydrogen-Council.pdf.

Chapter 2
The Future of Energy and Mobility

As introduced in Chap. 1, the full-scale commercialisation of hydrogen as an energy carrier for direct transport applications has not yet been achieved at scale despite the introduction of the phrase "hydrogen economy" more than 40 years ago. This reality may partly be due to: significant technical challenges; a lack of economic viability at technically accessible scales of production; the lack of an appropriate business model; and a lack of suitably long-term decision-making by large industrial concerns. Much emphasis has been given both to laboratory-based research and development and to policy-push in order to foster disruptive technological innovation, but so far no such large-scale systemic transition has occurred.

2.1 A Very Brief History of the Car

Arguably, the technology that had the greatest impact on the twentieth century was the private car, and throughout that period, one means of propulsion had total dominance—the internal combustion engine (ICE). The ICE was developed in Europe by pioneers such as Étienne Lenoir and Nikolaus Otto in the second half of the nineteenth century. Numerous inventors and engineers on both sides of the Atlantic contributed to the development of the technology such that by the dawn of the twentieth century, the ICE was ready to drive the world. The internal combustion engine of the twentieth century had the following key attributes:

- Liquid fuelled
- Long-ranged—hundreds of kilometres
- Rapidly filled—minutes
- Safe and reliable in operation.

The ICE achieved these capabilities via:

- Compressed gas intermittent combustion
- Use of a mechanical gearbox to change gear ratios.

© Springer Nature Switzerland AG 2020
W. J. Nuttall and A. T. Bakenne, *Fossil Fuel Hydrogen*,
https://doi.org/10.1007/978-3-030-30908-4_2

The mechanical gearbox was a triumph of the twentieth-century mechanical engineering initially via the development of synchromesh mechanical transmissions (first commercialised 1928) and later the emergence of fully automatic transmissions (commercialised from 1940 onwards). These technologies allowed the car to operate smoothly over a full range of speeds, including from stationery while utilising a rotating power source (the ICE) that cannot operate at low revolutions for risk of stalling. Automotive sector experts WardsAuto estimated that in 2010, the number of vehicles in operation worldwide had exceeded one billion for the first time [1]. The triumph of the internal combustion engine might appear complete.

A key part of the success of the internal combustion engine lies in the very high energy density of liquid hydrocarbon fossil fuels, such as gasoline and diesel. From these aspects come the key benefits in terms of range and rapid refuelling. They also, however, relate to the large amounts of gaseous emissions produced by combustion. In the case of these fuels that process brings with it the production of large amounts of carbon dioxide—a key greenhouse gas responsible for global climate change. The environmental drawbacks of gasoline and diesel are putting pressure on automotive manufacturers worldwide to make a profound shift. That shift is underway, and it is a shift to the electric vehicle power train. However, perhaps ironically the greatest single motivation for this transition came not from the need for global climate change mitigation but from a concern for urban air quality. The combustion of diesel fuel, in particular, is associated with the production on nitrogen oxides harmful to human health. The emission of these NOx pollutants is highly regulated, especially so in the USA where by 2015, the German headquartered automaker Volkswagen had gradually grown its market share largely on the strength of the slightly better greenhouse gas emissions properties of diesel fuel compared to gasoline. In the period from 2009 to 2012, Volkswagen saw its VW branded sales increase by 105% [2].

In 2015, however, Volkswagen hit a serious pothole on its American road-trip. The company was found to be achieving its passing results in US NO_x tests by means of "defeat devices". An example of such a vehicle on display in 2010 is shown in Fig. 2.1. While the legality of such an approach can be debated, in the court of public opinion the result was clear, this was cheating on a grand scale. The consequence for Volkswagen was to drive a volte-face in company strategy away from diesel power trains and towards potentially lower carbon (if green electricity can be sourced) electric vehicles.

Another significant force behind the development of electric cars has been the passion and enthusiasm of PayPal billionaire Elon Musk, co-founder of the Tesla Motor Company. While the company has struggled in its hoped-for journey to profitability, it has been very successful in popularising the notion of the electric vehicle as a high performance premium choice—starting from the initial Tesla Roadster based upon British Lotus bodywork and external design. Elon Musk saw early on that, while people might want a greener car, they also want a better car.

In 2018, it seems that all major car companies are adjusting towards the new electric reality. What does all this imply for the future of hydrogen energy?—This we shall consider in the next section.

Fig. 2.1 VW Golf TDI "clean diesel", Washington Auto Show USA 2010. *Source* Mario Roberto Duran Ortiz, Licence: CC BY-SA 3.0

2.2 Fuel Cell Electric Vehicles and the Future of Mobility

The future of hydrogen-fuelled mobility lies with the fuel cell. A fuel cell is similar to a battery in that it generates electricity from electrochemical reactions. The essential difference, however, relates to how energy is provided to the cell. In the case of a battery, both the energy-out and the energy-in take the form of electricity, but in the case of a fuel cell—while the useful energy-out is primarily electricity, the energy-in is provided by a fuel, such as hydrogen. As long as there is fuel, the cell will produce electricity. A key issue with fuel cells, however, is fuel purity. Different fuel cell technologies have different abilities to cope with variable fuel quality. Different types of fuel cell are summarised in Table 2.1. While noting that slow start-up is a problem for some fuel cell technologies when considering vehicle-based applications, fuel cell systems operating at high-temperatures could directly avoid a problem encountered in electric vehicles: heating the passenger compartment [22]. Electric vehicles face a problem owing to the high energy requirements associated with space heating. One possible means to mitigate the problem is to minimise the air volume, i.e. the passenger compartment size. Clearly, this is an unattractive choice for car designers. Generally, space heating is easier for all types of hydrogen-fuelled cars because the energy storage capabilities of hydrogen systems tends to exceed the storage potential of battery systems. Some hydrogen technologies also helpfully run hot during operation (see Table 2.1). Even when a low-temperature fuel cell is used, small amounts of hydrogen may easily and safely be diverted for vehicle space heating. The relative ease of space heating is one reason that hydrogen-powered vehicles can more easily resemble traditional vehicle shapes. Another attribute is that vehicles facing range difficulty (such as some battery electric vehicle concepts) are pushed towards weight reduction through design. The design pressures are generally rather different for hydrogen-powered vehicles. In time, hydrogen-powered vehicles will allow for the emergence of new vehicle design concepts.

Table 2.1 Summary of different fuel cell technologies [3]

Acronym	Full description	Fuel requirements	Operational temperature	Notes
DMFC	Direct methanol fuel cell	Methanol	Low temperature (near ambient temp—approx. 50–120 °C)	Not yet commercialised A variant of PEMFC Expected to find application in transport and mobility
PAFC	Phosphoric acid fuel cell	A broad range of fuels is possible, but if petroleum is used it must be desulphurised	Moderate temperature (approx. 150–200 °C)	First fuel cells to be commercialised Suitable for stationary applications
PEMFC	Proton-exchange membrane fuel cell	Hydrogen	Low temperature (ambient temp)	Not yet commercialised High efficiency of hydrogen conversion Expected to find application in both stationary and transport/mobility contexts
SOFC	Solid oxide fuel cell	Highly tolerant—including a wide range of fossil fuels	High temperature (in excess of 600 °C)	Solid oxide electrolyte—sometimes ceramic. Typically slow start-up and hence used predominantly in stationary applications

Fuel cell electric vehicles hold out the prospect of mobility and transport with some key beneficial attributes. Hydrogen is an environmentally benign gas with no direct greenhouse gas or ozone depletion concerns. Small secondary effects are, however, possible, and these have been reviewed in the literature [4].

The fuel cell converts hydrogen fuel to electricity by combining the hydrogen with oxygen to produce water. The only emission from the vehicle is pure water. Hydrogen can be transferred to the vehicle rapidly over timescales similar to today's petroleum refuelling (more about this later in this chapter). A car can be fuelled easily with sufficient energy to offer ranges similar to those of today's traditional vehicles. Overall, the FCEV approach offers the prospect of a very low "well to wheels" environmental impact (although, as this book emphasises, some significant attention must be paid to the source of the hydrogen used and the emissions associated with that activity).

Importantly, an FCEV future will be fuelled using infrastructures very similar to today's petroleum economy. Indeed, one can expect the fuel supply to be provided by today's international oil companies. As such, the FCEV approach represents a means by which today's oil and natural gas companies could transition to an ultra-low-carbon future. The proposition is one of incremental innovations rather than the disruptive displacement of incumbents. Given the real engineering and technical expertise of the incumbent fuel companies, the more evolutionary approach would appear to have much to commend it.

There are, of course, concerns around hydrogen safety and handling, but these are arguably small when compared to the challenges presented by gasoline; such issues will be considered later in this chapter.

Fundamentally, it is important to stress that a fuel cell electric vehicle is an electric vehicle. Many of the innovations and developments in vehicle technology associated by the recent enthusiasm for battery electric vehicles will be transferrable to later FCEV products. While the interest of major car manufacturers in battery electric vehicles (BEVs) is very visible and much publicised, there is also a widespread ambition in time to shift the technology across to hydrogen-fuelled FCEV power trains. This next step fits in with a strategic sequence of automotive technology development following a pattern involving a set of related innovations and technologies, namely:

Hybrid electric vehicle, HEV—Hydrocarbon-fuelled ICE with partial electric power train

Battery electric vehicle, BEV—Electricity charged—fully electric power train

Plug-in hybrid electric vehicle, PHEV—HEV with option of electric charging

Fuel cell electric vehicle, FCEV—Hydrogen fuelled—fully electric power train.

Generally when compared to ICE vehicles, hydrogen-based fuel cells bring a set of design advantages including high electrical efficiency, zero harmful tailpipe emissions, reduced cabin noise and excellent reliability arising in part from the lack of moving parts in the power train.

Looking further ahead, one can continue to ask how passenger car electrification (BEV and FCEV) will adjust vehicle design and car body shapes, in particular. All internal combustion engines must find space for an engine block and a power transmission gearbox. These heavy metal items affect the handling and stability of the car as a whole. Looking ahead battery systems or fuel cell stacks offer the prospect of more flexibility in terms of shape and in-vehicle positioning. An FCEV, however, brings with it the requirement for a hydrogen storage tank. One interesting emerging development is for electric motors mounted entirely within the structures of vehicle wheels. Such developments are sometimes known as "hub motors". Various companies around the world are competing to develop a commercial offering able to propel a car with no detriment to its handling and ease of use. Key to this is that the weight and inertial properties of the new wheels should feel and handle similar to traditional wheels. Already prototypes are being tested with powers in the range

50–100 kW. Once established, this development has the potential to radically alter car design freeing up much space in the heart of the vehicle and enormously reducing the parts of the car where mechanical moving parts are needed. We suggest that this is a technology to watch.

More generally, Toyota has been a world leader in driving forward new vehicle technology developments. The Toyota Prius established Hybrid Electric Vehicle technology worldwide. Even this success has been slow. The Toyota Prius required 13 years to capture roughly 2.3% of the US car market following its US launch in 2000 [5]. Furthermore, Toyota has also led the way into hydrogen-based FCEV technology having produced the first commercially available hydrogen car—the Toyota Mirai.

2.2.1 The Toyota Mirai

For the last decade, Toyota's attention has shifted towards hydrogen. The Mirai is an electric vehicle that uses hydrogen fuel cell technology in combination with a storage battery. Hydrogen reacts within the fuel cell to produce electricity. This electricity charges the battery, together powering a motor that drives the car. Within the fuel cell, the hydrogen combines with oxygen to produce water, and that is the Mirai's only tailpipe emission.

In 2015, the Mirai became the first commercially available hydrogen fuel cell electric vehicle (FCEV). The Mirai electrolyte fuel cell stacks can produce 114 kW, connected to an electric motor and a battery. The fuel tank can hold around 5 kg of hydrogen, which gives it a range of 310 miles [6, 7].

Toyota hopes that the Mirai could be for hydrogen fuel cell vehicles what the Prius was to hybrids. The main obstacle to market penetration, aside from the immature state of the hydrogen market itself, is the time and effort it takes to make the fuel cell stack and hydrogen tanks. Tolerances are critical in such matters. Hydrogen risks are mitigated via the careful engineering of pressurised storage systems, and every set of twin tanks is tested with inert helium to ensure there are no leaks from the slim pipes feeding into the stack [8] (Fig. 2.2).

There are two main advantages of hydrogen fuel cell power trains over battery electric vehicle (BEV) alternatives. It takes just a few minutes to fill a hydrogen tank, whereas it takes much longer to charge a BEV even at a rapid-charge powering station. While hydrogen production has the potential to be generated from low-carbon sources, presently 78% of hydrogen for mobility is still produced from natural gas and oil in ways that do nothing to abate harmful greenhouse gas emissions [9]. Such harmful emissions must be avoided, either via a transition to "Green Hydrogen", or as this book argues, via a cleaning-up of Mature Hydrogen production methods.

Similar to petroleum and liquefied petroleum gas (LPG), hydrogen can be stored relatively easily and safely and can be sold at the same fuelling stations as these existing products. However, unlike fossil fuel tanks, hydrogen storage tanks cannot be buried beneath the forecourt of the filling station. Hydrogen storage facilities must be installed above ground. All the attempts to commercialise the production, storage

Fig. 2.2 The Toyota Mirai commercially available hydrogen fuel cell electric car, Zaragoza, Spain, June 2016. *Photo* Authors (WJN)

and transport of hydrogen are challenged by the evolution of battery electric vehicles (BEV). The charging of BEVs at home and at work challenges the business models of the international oil companies (IOCs). IOCs are offering rapid recharge services, especially at highway locations, but most BEVs struggle to accept multiple rapid recharges in quick succession. As such, the focus for BEVs has to this point mostly been on urban mobility applications as opposed to trains, heavy trucks and buses.

Hydrogen fuel cells do not suffer from the main BEV drawback of limited range. Figure 2.3 shows FCEVs can cover longer distance compared with BEVs. Furthermore, once the hydrogen is depleted, an FCEV can be refuelled in a matter of minutes,

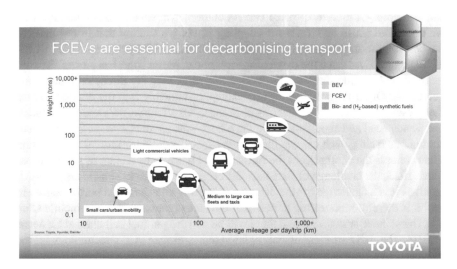

Fig. 2.3 Optimal decarbonised fuel options for different vehicle types. *Original Source* Toyota, via [6]

similar to a petroleum vehicle. In a country with a less developed fast-charging infrastructure, the BEV option will be a much harder sell for drivers contemplating longer journeys compared to FCEV. Importantly, the FCEV approach is well suited to the handling of heavy loads, hence the interest in bus and truck applications.

The success of the FCEV approach depends on cheaper low-carbon hydrogen production and a drive by the energy companies to increase hydrogen retail supply infrastructure investment. In order to attract this investment, Toyota has embarked on promoting their FCEV by focussing on leasing the Mirai to selected organisations (taxi operating companies, police forces, etc.) rather than to individuals. The aim is to identify user communities likely to need very high levels of resource utilisation and to be able to benefit from fleet services. It should be stressed, however, that unlike hydrogen fuel cell vehicles from Honda and Hyundai, which exist only within tightly controlled lease programmes, the Toyota Mirai can be purchased outright as well as leased. The Toyota Mirai is, however, not an inexpensive choice. Autocar magazine reports the price of a UK Toyota Mirai in 2018 as being £65,219 [10]. Despite such realities, Toyota's target for 2020, and beyond, is to have produced 30,000 FCEV for the global market [6].

2.2.2   Green Car and Better Car

Despite the troubles that Elon Musk's Tesla Inc. has had on the hoped-for journey to financial success, Musk's vision for the roll-out of electric vehicle technology has been noteworthy for its approach. His first product was a small two-seater roadster sports car based upon the British Lotus Elise, but with an entirely novel Tesla power train based upon lithium-ion batteries. Since then the company has pushed into luxury vehicles with the model S. Tesla Inc. took the view that a new car must be a better car, not simply a more ecological car. As authors sympathetic to such a approach we commend Musk and Tesla for adopting it, but we caution that in our opinion the best long-term prospects for such an approach lie with hydrogen-fuelled FCEV technologies. Ever since the dawn of the motor car, motor sport has pushed forward innovation. Examples include disc brakes and more recently direct shift automatic gearboxes. Of all the many types of motorsport, arguably it is the endurance race that presents the greatest engineering challenges. Of all the world's endurance races, one stands head and shoulders above the others—the annual Le Mans 24. In 24 h of continuous racing, each car can these days expect to drive for more than 5000 km. The winner is the car that goes the furthest in 24 h driven by a team of three drivers. Since 1923, the challenge of Le Mans has been a challenge to vehicle engineers as much as it is a challenge for the drivers—reliability and sustained performance are key. It is therefore interesting to note that the race organisers, the Automobile Club de l'Ouest, have announced that there will be a hydrogen fuel cell class at Le Mans in 2024. The ACO are clear in their opinion that hydrogen FCEV technology is "the next objective" in decarbonising motorsport [11]. In June 2016, a 544-bhp hydrogen-fuelled car was seen driving the Le Mans circuit ahead of the famous race [12]. The car was the "Green GT H2" driven by former Formula 1 driver Olivier Panis. While some might argue that such initiatives are merely an attempt to "green-wash" the fossil fuel frenzy that is Le Mans, we take the view that this first step was indeed a vision of the future, not just for Le Mans and for motorsport, but arguably much more generally.

2.3 Hydrogen Safety

One of the first issues to be raised in any discussion of hydrogen is the matter of safety. One point to remember is that the traditional alternative, petroleum, is far from hazard-free, indeed in key respects hydrogen is less hazardous. Petroleum fuels can flow and splash, whereas hydrogen fires burn vertically (see Fig. 2.4). Pure hydrogen fires are, however, almost invisible—first responders might only see a shimmer in the air from heat effects. Also hydrogen fires produce somewhat less heat than petroleum fires. Hydrogen fire safety issues have been well summarised by Ricci et al. [13]

A key issue of hydrogen safety is (for pure hydrogen) its odourless dispersion in air. Hydrogen air mixes present a serious explosion risk. For this reason, much concern for hydrogen safety relates to enclosed volumes (such as vehicle cabins and closed domestic garages). Risks of gaseous H_2 leakage into confined spaces are a key concern as detonation can occur for mixes in the range from 18.3 to 59% by volume. The low viscosity of liquid hydrogen and for both gaseous and liquid hydrogen, the small molecular size and the low molecular mass lead to serious leakage risks. As a consequence, there is significant expert interest in ensuring hydrogen safety (Fig. 2.5).

The fundamental consideration in hydrogen safety design is the prevention of the release of H_2 molecules. One technique that can prove valuable is gaseous nitrogen jacketing. In addition the controlled oxidation of any leaks can help reduce the risk

Fig. 2.4 Green GT H2 Hydrogen-fuelled endurance racer demonstrator. *Source* Autocar [12]

Fig. 2.5 Equivalent vehicle fires at 3 and 60 s after ignition (H_2 flow rate: 5 kg/min) the left-hand images show hydrogen and the right-hand images gasoline. *Source* M. R. Swain [14]

of the uncontrolled release of H_2 molecules. Hydrogen detection and alarm systems can add to safety, but they should be designed and implemented so as to respond quickly.

Noting the fact that genuine hydrogen safety concerns exist, there is also a somewhat separate and very real issue—public anxiety. It has long been known that fear and danger are not well-aligned concepts. While of course there is a tendency to be frightened by genuine danger, it is not the case that the greatest fear correlates with the greatest danger. Other important psychological factors intervene when considering the determinants of fear. Such matters have been considered extensively by Paul Slovic, Professor of Psychology at the University of Oregon, USA. For example, activities that are voluntary are found to be less frightening than activities that are mandated even when the danger is the same. When considering anxieties around hydrogen, two things deserve some particular consideration: first, as discussed, the real and special dangers and second items that may have entered the popular consciousness such as the 6 May 1937 Hindenburg airship disaster in which 36 people died and which was memorably captured by news reels and radio reporters. Only years later did it become apparent that much of the fire risk related more to the doped envelope of the airship than from the hydrogen gas inside. By then, however, the association of hydrogen with fiery disaster was already made.

2.4   Vehicle Refuelling

Particularly since the diesel emissions scandal that became public in 2015, battery electric vehicles have received an upsurge in industrial interest. The technology, however, continues to suffer from some serious disadvantages when compared to traditional gasoline-fuelled vehicles (including hybrids) and also emerging hydrogen technologies. Battery electric vehicle technologies are associated with slow recharging times. The once-mooted idea of rapid battery swaps has receded in the face of concerns relating to the very high value of batteries, property rights and the complexities of leasing and similar business model issues. The emerging consensus for electric vehicles is based around slow charging overnight at home or in the daytime at work with the option of so-called rapid charging at filling stations or public access points. This option has, however, come in for some criticism in the UK. While rapid charging can be very effective mid-journey, drivers of the Nissan Leaf BEV have reported that follow-on attempts are highly problematic with the second attempt taking far longer than the usual 40 min associated with rapid charging [15].

As noted earlier, range and the time required for recharge remain a serious concern, especially for lower-cost battery electric vehicles. That said, however, plug-in BEV technology currently gives users access to a very low-tax energy option, namely domestic electricity. While even the lowest cost electric vehicles are significantly more expensive to buy than their direct conventional (fossil fuelled) equivalents, the cost of energy (electricity) is dramatically lower, in Europe at least, largely as a consequence of the high taxes and duties imposed on fossil fuels for transport.

One difficulty limiting the role out of BEV technology is that in urban areas, relatively few people have access to private off-road parking with easy access to their home electricity supply. It is simply not legal, or appropriate, to run power cables across public sidewalks and footpaths. These problems are most acute in cities with very old infrastructure and layouts (such as many European urban centres). Many people live in apartments converted from nineteenth-century large town houses. How is an upper-floor apartment dweller supposed to charge their electric car? The usual answer would be to expect the local authority to install public access charging points throughout the city and especially in residential areas. This may indeed occur, but many parts of local government are still coping with significant austerity issues arising from the fallout from the 2008 financial crisis.

In contrast to the difficulties faced by the roll-out of viable BEV charging, hydrogen vehicles can continue the well-established practices associated with traditional petroleum-based refuelling. Hydrogen could retain the role played in retail energy supply by the international oil companies (Shell, BP and Exxon etc.).

Generally, the quantity (by mass) of hydrogen required by an FCEV will be lower than the quantity of petroleum required for an equivalent internal combustion engine. A typical modern car has a range of 400 km requiring roughly 24 kg of petroleum for its combustion engine. For equivalent range with a hydrogen-fuelled FCEV, just 4 kg of fuel would be required. While, of course, this mass of hydrogen might appear to be impressively low, one should not forget the relatively low mass density of liquid hydrogen and the very low mass density of gas even at high compression. As such, it is not the mass of the fuel, but rather the bulkiness of the storage tank that must be managed by FCEV designers (Fig. 2.6).

As a consequence of the safety risks described earlier, the refuelling of hydrogen-fuelled vehicles is something that must be approached carefully. The more common approach is high-pressure gas fuelling at either 700 or 300 bar. Liquid hydrogen refuelling systems have also been developed by Linde and BMW. The high-pressure gas fuelling approaches (at both 700 and 300 bar) are now technologically mature

Fig. 2.6 The Hyundai ix35 Fuel cell Electric Vehicle. This car is fuelled with 700 bar hydrogen gas. It is photographed at the SINTERF Headquarters in Oslo. *Source* Bartek Glowacki)

and have been installed in commercial settings. The cryogenic liquid hydrogen technology, however, involves more advanced engineering. This liquid hydrogen transfer technology has been patented and is ready for commercial roll-out via the licensing of the intellectual property involved.

Liquid hydrogen can feature at the filling station in one of two ways. For example, it can be used simply as a local hydrogen storage system at the station, and perhaps associated with the upstream supply chain. In that case, the supply to vehicles might simply be in the form of high-pressure gas at either 700 or 300 bar. The more ambitious proposition is that developed by Linde and BMW whereby liquid hydrogen is actually supplied to the vehicle. In that case, the vehicle must be equipped with an appropriate cryogenic storage tank. In recent years, second-generation cryo-compressed hydrogen vehicle fuel tanks have been developed. These new technologies are associated with very low boil-off and provide improved security against risks associated with boil-off of the last cryogenic hydrogen liquid. In this event, the tank can continue as a high-pressure gas tank supplying hydrogen at 300 bar. BMW has summarised the status of hydrogen storage technologies for on-vehicle applications; see Fig. 2.7.

2.5 Low Hanging Fruit

Despite all the enthusiasm for hydrogen-fuelled FCEV cars, it is not the privately owned family car that will prove to be the most attractive first application of these innovative technologies. Currently, the most attractive sectors to adopt hydrogen-based technologies will be long-distance road haulage and urban bus fleet operators

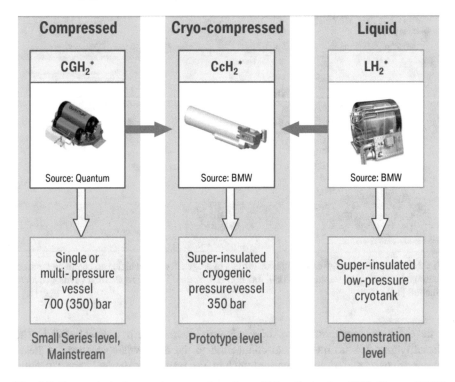

Fig. 2.7 Cryogenic hydrogen storage options for vehicles, September 2012 (*Source* BMW [16]). Acronym key: First column: CGH2—compressed gaseous hydrogen 700 bar; second column: CcH2—cryo-compressed hydrogen in the range 10–350 bar; and third column: LH2—liquid/liquefied hydrogen (atmospheric pressure - ca. 10 bar)

(attracted by the low on-vehicle emissions and operational advantages over battery electric vehicles); see also Fig. 2.3. Large trucks on the world's highways typically travel such distances and pull such heavy loads that battery-based options are not satisfactory. The issues around various types of vehicle and their applications are further summarised and extended by Fig. 2.8.

While Tesla, Nissan and others have already revealed the viability of battery electric vehicles for passenger cars (although as discussed earlier some issues do still remain), it is in the area of heavyweight and long-distance transport that BEV technology clearly falls short. The shift to FCEV commercialisation is being seen first in larger road vehicles such as buses and heavy trucks and even in trains. Before returning to the issues of road transport, let us briefly pause to consider the role of hydrogen in future rail systems.

In many parts of the world, rail systems are already electrified. Such electrification is an expensive infrastructure choice appropriate only for lines operated with high levels of utilisation. Many rail lines are, however, only used sparsely and until recently the only options have been fossil fuelled. It should be noted in passing that the concept of the hybrid electric vehicle (such as a Toyota PRIUS) takes partial inspiration from

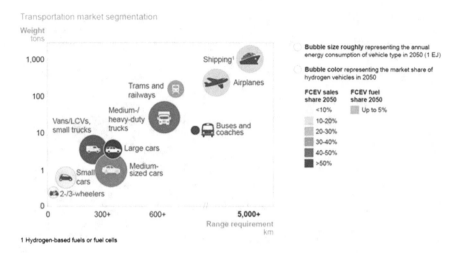

Fig. 2.8 Weight and range capabilities of various hydrogen-fuelled vehicles and predicted market share and energy consumption in 2050, c.f. Fig. 2.3. *Source* Hydrogen Council (hydrogencouncil.com) [17]

a far older rail innovation—the diesel-electric locomotive. In a diesel-electric, the diesel engine has no mechanical connection to the drive wheels rather the diesel motor drives an electricity generator which in turn provides electrical power (usually as direct current) to electric motors attached to the drive wheels. Such a power train is enormously simpler and hence easier to maintain than a more traditional diesel motor-based drive train. In a hybrid car, of course, the concept is extended with the addition of electricity storage batteries. Indeed, that aspect is now being taken up with hybrid diesel-electric locomotives which bring in the beneficial possibility of regenerative braking.

With the role of electric drive well established in trains and locomotives what can be done in respect of those rail routes that are poorly suited to electrification using overhead wires or third rails? The challenge of system decarbonisation is presenting a growing challenge, and the limited energy capacity of onboard battery storage precludes such an approach for a vehicle as heavy as a freight train, or even a passenger train over a long journey. One clear solution, environmentally preferable to the hybrid diesel electric train mentioned earlier would be a hydrogen fuel cell powered train. Such trains, as built by the French company Alsthom, are now operating over a 100-km route in northern Germany [18]. The two Coradia iLint passenger trains can run for 1000 km with a single charge of hydrogen fuel. Alsthom is now working with Eversholt Rail to bring similar technology to the UK [19]. The intention is to convert existing Class 321 electric trains to hydrogen fuel via the fitting of fuel cells and hydrogen tanks. Such plans are consistent with the UK government's environmentally motivated aspiration that diesel rolling stock will all be gone by 2040.

Similar concerns are motivating interest in hydrogen-fuelled systems for long-distance road haulage. In the USA, the Nikola Motor Company has launched the

Nikola One hydrogen-fuelled truck offering a range up to 1000 miles with a 20-minute refuel time [20]. The trucks are offered to haulage companies under lease arrangements, and an early high-profile customer has been the American brewing company Anheuser-Busch, famous for Budweiser beer, with a spring 2018 announcement of a deal involving up to 800 of the new trucks [21]. Nikola Motors takes its name from the famous Serbian-American electrical engineer, Nikola Tesla. The company faces competition from the inventor's other namesake company Tesla Inc. That company, famous for its BEV cars, announced on 16 November 2017 the intention to produce a wholly battery electric large truck with the first vehicles being ready in 2019. This proposition runs counter to much expert opinion concerning the viability of BEV technology in such applications. The Korean automaker Hyundai has announced its own hydrogen FCEV truck ambitions with a major deal with Zurich, Switzerland-based company H2 Energy [22]. The deal will supply up to 1000 trucks with a range of 250 miles and a refuel time of seven minutes. The arrangements will start in 2019 and endure for 5 years.

Around the world, there is much interest in hydrogen-powered buses in public transport systems (see Fig. 1.4). Many Chinese cities face particular problems with urban air quality. There is much effort in that country directed towards cleaner vehicle solutions. For example, it has been reported in August 2018 that Shandong Heavy Industry has plans to deploy 2000 hydrogen buses in Shandong province [23]. It appears that a global move towards hydrogen-fuelled trucks and buses is already underway. The high levels of stored energy capacity combined with rapid refuelling and the benefits of an emissions-free electric power train combine to motivate the shift away from diesel fuel. In closing this paragraph, it is worth mentioning another early hydrogen opportunity already well established is the adoption of hydrogen-fuelled fork-lift trucks in warehouse environments.

In this chapter, we have focussed on hydrogen utilisation in fuel cell electric vehicles. Indeed, for the reasons described around the rapid progress in electric vehicles of all types, it seems likely that is indeed that dominant way ahead for hydrogen in transport and mobility. Before closing the chapter, however, it is worth mentioning the use of hydrogen in internal combustion engines. BMW invested much effort in such ideas in the first decade of the twenty-first century. One hundred Hydrogen-7 cars were produced based on the standard gasoline 760Li model in the period 2005–2007. The proposition has been described by BMW [24]. A nice benefit when compared with the significant issues of "range-anxiety" surrounding Battery Electric vehicles was that the Hydrogen-7 was a dual-fuel vehicle (with two tanks) able to run on either gasoline or hydrogen depending on the ignition timings as set by the engine management system. BMW put significant effort into minimising and mitigating safety risks associated with hydrogen leakage. The plans also involved the use of a cryogenic tank holding 8 kg of liquid hydrogen. The whole Hydrogen-7 enterprise was, however, abandoned by BMW. It is reported that this followed a critical assessment by the US Environmental Protection Agency which ruled that even in hydrogen mode the vehicle could not be said to be zero carbon because of the use of oil-based lubricants (with some inevitable combustion of those molecules) [25]. Hydrogen also presents particular difficulties concerning combustion control.

We understand that BMW found that in hydrogen-fuelled operation, their engines were significantly damaged after just 10,000 km of operations, a level far too low for a viable commercial proposition. The difficulties of engine management are described in a paper by Wolfram Enke and colleagues [26].

Despite the difficulties encountered by BMW with the Hydrogen-7, interest in hydrogen-fuelled internal combustion continues. In the UK a small company OakTek has developed the Pulse-R, essentially to replace the cylinder head of conventional four-stroke gasoline-fuelled piston engines [27]. It can cope well with a wide range of fuels including hydrogen. The approach involves very lean combustion and avoids the production of harmful nitrogen oxides [27].

Another avenue involves a linkage between hydrogen as a fuel for the future and the Wankel rotary engine—a concept of beautiful engineering simplicity that nevertheless failed to displace the established piston engine despite much investment and support particularly by Mazda in Japan. The attachment of Mazda to the rotary engine is hard to overstate, and in 2015, it received attention from Wired magazine [28]. Mazda has recently embarked on a plan to bring back the rotary engine in a hydrogen-fuelled form. The last of Mazda's original gasoline fuelled concepts, the RV8, had been forced out of production in 2012 by ever tightening emissions reduction requirements. The possibility of hydrogen fuels would allow Mazda to return to the company's first love—the rotary engine.

Acknowledgements We gratefully acknowledge the assistance of Professor Bartek Glowacki. All responsibility for errors or omissions rests with the authors.

References

1. Sousanis, J. *World vehicle population tops 1 billion units*. August 15, 2011 [cited September 21, 2018]. Available from: https://www.wardsauto.com/news-analysis/world-vehicle-population-tops-1-billion-units.
2. Brennecke, H. 2013. *Volkswagen in the United States: An evolving growth story*. January 16, 2013. Available from: https://www.volkswagenag.com/presence/investorrelation/publications/presentations/2013/01-january/2013-01-16+DTW+DB+Conf+Website+Top+Copy.pdf.
3. Nuttall, W.J., B.A. Glowacki, and S. Krishnamurthy. 2016. Next steps for hydrogen. London: Institute of Physics.
4. Derwent, R., et al. 2006. *Global environmental impacts of the hydrogen economy*, vol. 1, 57–67.
5. Total US car sales roughly 6.3 million (2017) [Source: Statista] Peak Toyota Prius sales (2013) 145,000 [Source: http://www.hybridcars.com/is-toyotas-king-of-hybrid-sales-the-prius-losing-its-edge/]. Hence Toyota Prius achieved roughly 2.3% of the US market. See also: *Worldwide sales of Toyota motor hybrids top 4 M units - Prius family accounts for almost 72%*. Energy, technologies, issues and policies for sustainable mobility, 22 May 2012; Available from: http://www.greencarcongress.com/2012/05/tmc-20120522.html.
6. Cooper, D. 2018. *It's too early to write off hydrogen vehicles*. Available from: https://www.engadget.com/2018/05/29/hydrogen-fuel-cell-toyota-mirai-evs/?guccounter=1.
7. Hasegawa, T., et al. 2016. *Development of the fuel cell system in the Mirai FCV*, 5. SAE International.

8. Toyota. 2018. *The facts.* Available from: https://www.toyota.co.uk/new-cars/new-mirai/the-facts.
9. Toyota. 2018. *Company background.* Available from: http://media.toyota.co.uk/wp-content/files_mf/1526917964180412MCompanybackground.pdf.
10. Autocar. 2018. *Toyota Mirai prices and specs.* October 8, 2018. Available from: https://www.autocar.co.uk/car-review/toyota/mirai/specs.
11. McKenzie Smith, P. 2018. Hydrogen is the future of decarbonised mobility—Fuel cell endurance cars confirmed for Le Mans. *The Telegraph.* London.
12. Burt, M. 2016. *Green GT H2 Hydrogen racing car makes UK debut at Goodwood.* Available from: https://www.autocar.co.uk/car-news/motorsport/green-gt-h2-hydrogen-racing-car-makes-uk-debut-goodwood.
13. Ricci, M., P. Bellaby, and R. Flynn. 2010. Hydrogen risks: A critical analysis of expert knowledge and expectations. In *Hydrogen energy: Economic and social challenges*, ed. P. Ekins, 217–240. London.
14. M.R. Swain. *Fuel leak simulation.* University of Miami, undated. Available from: http://evworld.com/library/Swainh2vgasVideo.pdf.
15. Milligan, B. *Electric car buyers claim they were misled by Nissan.* BBC Business 2018, June 29, 2018. Available from: https://www.bbc.co.uk/news/business-44575399.
16. Kunze, K., and Kircher O. 2012. Cryo-compressed hydrogen storage. Presentation at Cryogenic Cluster Day, Oxford, September 28, 2012. Available at: https://stfc.ukri.org/stfc/cache/file/F45B669C-73BF-495B-B843DCDF50E8B5A5.pdf.
17. Hydrogen scaling up-a sustainable pathway for the global energy transition. 2017. Available from: http://hydrogencouncil.com/wp-content/uploads/2017/11/Hydrogen-scaling-up-Hydrogen-Council.pdf.
18. Agence-France Presse, Germany launches world's first hydrogen-powered train, as published in The Guardian, September 17, 2018. Available at: https://www.theguardian.com/environment/2018/sep/17/germany-launches-worlds-first-hydrogen-powered-train.
19. Fuel cell trains 'on their way to the UK', 11. Energy World, June 2018.
20. Nikola Motor Company, Nikola One website information: https://nikolamotor.com/one.
21. Johnson, Eric M. 2018. Anheuser-Busch orders up to 800 hydrogen-fueled big rigs. Reuters Technology News, May 3, 2018. Available at: https://uk.reuters.com/article/us-ab-inbev-orders-nikola/anheuser-busch-orders-up-to-800-hydrogen-fueled-big-rigs-idUKKBN1I41JI.
22. Ed Wiseman.2018. Hyundai to supply 1000 hydrogen fuel cell lorries in Switzerland, Daily Telegraph, September 27, 2018. Available at: https://www.telegraph.co.uk/cars/news/hyundai-supply-1000-hydrogen-fuel-cell-lorries-switzerland/.
23. China: Shandong Heavy Industry to Develop and Implement 2,000 Hydrogen Fuel Cell Buses in Shandong Province, FuelCellsWorks, August 24, 2018. Available at: https://fuelcellsworks.com/news/china-shandong-heavy-industry-to-develop-and-implement-2000-hydrogen-fuel-cell-buses-in-shandong-p.
24. BMW Hydrogen 7. November 2006. Available from: https://www.wired.com/images_blogs/autopia/files/bmw_hydrogen_7.pdf.
25. Nica, G. 2016. *Why did BMW really stop making the hydrogen 7 model.* 7-series 2016, August 17, 2016. Available from: http://www.bmwblog.com/2016/08/17/bmw-stop-making-hydrogen-7-model/.
26. Enke, W., et al. 2007. The Bi-fuel V12 engine of the new BMW Hydrogen 7. *MTZ Worldwide* 68 (6): 6–9.
27. Andrew, P. 2017. What is the Pulse-R engine? Available from: https://www.oaktec.net/what-is-the-pulse-r-engine/.
28. Stecher, N. 2015. Mazda's confusing plan to resurrect the famously dirty rotary engine. December 1, 2015. Available from: https://www.wired.com/2015/12/mazdas-confusing-plan-to-resurrect-the-famously-dirty-rotary-engine/.

Chapter 3
Hydrogen Chemical Engineering—*The Future*

This book is devoted to the message that hydrogen has a beneficial role to play and that furthermore that role might be met by evolving today's hydrogen production from fossil fuels, such as natural gas.

In Chap. 1, we introduced the idea that natural gas is regarded by some as a transition fuel on a journey to a renewables-based energy system. In this book, however, we propose an alternative opinion that natural gas, in part via hydrogen, may have a more enduring role in the twenty-first century and beyond. Even those that suggest that natural gas has a role as a transition fuel accept that its contribution will be relevant for several decades. In part, the question relates to long-term role of renewable energy in our whole energy system, noting the rapid progress that has been made by renewable electricity production in recent years (Fig. 3.1).

Today, the production of hydrogen is dominated by a set of methods collectively known as "thermal processes" of which steam methane reforming is the most common. The two established industrial techniques of steam reforming and partial oxidation are closely related especially in the initial chemical processing steps. In both cases, the fossil fuel source is oxidised although in steam methane reforming the oxidation process is taken further. A key difference between the two techniques is that partial oxidation gives rise to a net output of energy, i.e. heat. As such, it is said to be "exothermic". For the process as a whole, steam reforming is "endothermic"—i.e. external heat must be supplied.

While we posit that such methods may in future have much to offer low carbon mobility, we must concede that at the time of writing the only hydrogen-related CO_2 capture in the world is done by Air Products at Port Arthur near Houston, Texas.

© Springer Nature Switzerland AG 2020
W. J. Nuttall and A. T. Bakenne, *Fossil Fuel Hydrogen*,
https://doi.org/10.1007/978-3-030-30908-4_3

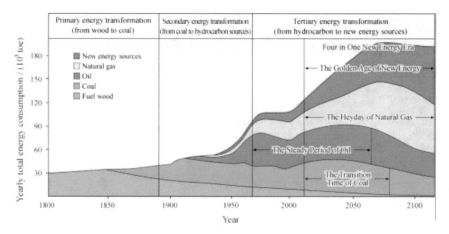

Fig. 3.1 Many authors point to the potential role of natural gas as a transition fuel from a declining fossil fuel energy system to a more renewables oriented system. This illustrative example is from Zou et al. These authors use the term "New Energy" to include both renewable energy and nuclear energy. *Source* Elsevier with permission [1]

3.1 Steam Methane Reforming

Steam methane reforming (SMR) is a well-established process of producing hydrogen at a commercial scale. Around the world, much investment continues to be placed in this catalytic approach, and there are potential good synergies in this technical approach with possible future options in carbon management, such as carbon capture and storage (CCS).

Steam methane reforming is a high-temperature process, typically requiring temperatures above 700 °C. The SMR process relies on catalysts, and traditionally, these have been nickel-based. For example, Johnson Matthey is a supplier of SMR catalysts including their HiFuel R110 pellets ($NiO/CaAl_2O_4$).

In the SMR process, steam reacts catalytically with methane:

$$CH_4 + H_2O \rightarrow CO + 3H_2 \tag{3.1}$$

It is a commonplace to combine the SMR reaction with a water-gas shift reaction to convert the carbon monoxide to less toxic carbon dioxide and to further enhance hydrogen production (Figs. 3.2 and 3.3):

$$CO + H_2O \rightarrow CO_2 + H_2 \tag{3.2}$$

A leading academic expert in the range of methods associated with the production of Mature Hydrogen and means by which associated greenhouse gas emission might be reduced is Nazim Muradov of the University of Central Florida, USA. In May 2017, he published a comprehensive technical review of the issues associated with

Fig. 3.2 Industrial steam methane reformer. *Source* Air Liquide; see: https://www.airliquide.com/media/air-liquide-starts-large-hydrogen-production-unit-germany

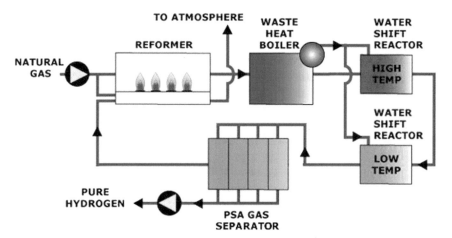

Fig. 3.3 Schematic of an industrial steam methane reformer including a waste heat boiler and water-gas shift reactors to convert carbon monoxide to carbon dioxide. In this schematic, the main vented gas includes carbon dioxide [2]

the production of hydrogen from fossil fuels with low or near-zero CO_2 emissions [3]. He observes that a modern SMR plant system, comprising an SMR unit, a water-gas shift reactor (WGS) and a Pressure Swing Adsorption Unit (PSA) can be expected to produce hydrogen to 99.999% purity. He goes on to note that about 60% of the total CO_2 produced is associated with the chemical process (i.e. the shifted syngas after the WGS) and the remaining 40% is the product of fuel gas combustion necessary to sustain an endothermic process. It is important to think of these two CO_2 sources separately. The process CO_2 can these days be generated at roughly 45% purity. This traditional waste stream must not be confused with the very dilute (maximum 11%) concentrations seen in fossil-fuelled combustion systems, and indeed in the flue gas of the SMR process itself. The relatively high purity of the process stream CO_2 renders it preferentially suitable for carbon capture and storage or even utilisation (see Chap. 5). If an alternative source of process heat could be identified (Muradov discusses high-temperature nuclear reactors and solar thermal systems as possible sources), then the flue gas emissions from SMR systems could also be largely eliminated. In these ways, the SMR process is amenable to significant emissions reductions.

The global dominance of the SMR approach is explained by a long-standing legacy of industrial experience. The SMR process was developed commercially in the early 1930s [4]. The favourable economics of this type of hydrogen production drove its growth and deployment from the start. In our previous work, we have looked in detail at the economics of hydrogen production and collated a range of third-party assessments [5]. In that work, a table is presented providing eight cost estimates for the production of hydrogen by electrolysis of water using electricity from renewable wind energy. These estimates of production cost range from US \$2.85/kg H_2 (itself the mid-point of an estimated range) to US \$7.3/kg H_2. **The average (mean) of these electrolysis-based estimates is US \$4.8/kg H_2**. Similarly, five estimates are provided for the production of hydrogen by SMR using natural gas feedstock and without CCS. These estimates range from US \$0.55/kg H_2 to US \$2.04/kg H_2 (itself the mid-point of a range). **The average (mean) cost for production by SMR is US \$1.3/kg H_2** (for this approximate calculation, we neglect to correct for the fact that the various estimates reported are given for different years ranging from 2004 to 2020). One thing appears clear, however, as things stand today, on average (and taking an inclusive view of total costs) SMR production is amongst the lowest cost hydrogen production processes available.

Autothermal reforming is a variant of the SMR process. It comprises partial oxidation followed by catalytic reforming. When deployed for natural gas feedstock and when using carbon dioxide (as opposed to steam), the relevant chemical reaction is:

$$2CH_4 + O_2 + CO_2 \rightarrow 3H_2 + 3CO + H_2O \tag{3.3}$$

Nazim Muradov has observed that oxygen-blown autothermal reforming is currently used in very large units for the production of syngas for Fischer–Tropsch or for methanol synthesis from a natural gas precursor. A key benefit of the ATR method is that CO_2 is generated at a pressure of three atmospheres facilitating the process [3].

3.2   Partial Oxidation Gasification

Although the majority of hydrogen is made via SMR processes, as described above, it is important to stress the importance of facilities based on principles of partial oxidation.

The POX process oxidises the fossil fuel as far as carbon monoxide and hydrogen. If the source fossil fuel were methane (the main constituent of natural gas), then the chemical reaction would correspond to Eq. 3.1. As noted earlier, the process is exothermic with operating temperatures as high as 1200 °C, but if catalysts are used this can be reduced to around 850 °C [6]. The lack of any need for external heating allows for POX reactors to be relatively compact, but the relatively high temperatures of operation lead to higher heat losses, and with limited process benefit to be gained from heat recovery, efficiencies (70–80%) are slightly below those seen in steam reforming [6]. "HYCO" plants are POX systems co-generating two important molecular products: hydrogen and carbon monoxide.

POX gasification does not shift the carbon monoxide through subsequent oxidation to carbon dioxide exhaust gas (see Eq. 3.2). For the HYCO POX process to be economic, it is important that both chemical products (carbon monoxide and hydrogen) have a local buyer. As a consequence, HYCO plants operate in the relatively few regions of the world with highly developed chemical process industries requiring both hydrogen and carbon monoxide. In these places, the carbon monoxide is an industrial feedstock gas. One example of such a facility is the Air Products HYCO III POX plant in La Porte Texas. This installation will be described in more detail later, in Chap. 6.

3.3   Innovation in Hydrogen Production

Given the industrial maturity of both the steam reforming and POX processes, it is easy to subscribe to the view that there is little potential for significant innovation in such methods of hydrogen production. In this book, however, we would like to propose a different view.

In Table 3.1, we refer to the possibility of plasma-based hydrogen production. Plasma processing is an example of a relatively new and innovative approach to hydrogen production, but even with the well-established thermal production techniques, further innovation is possible. For example, the catalysts used are in many cases proprietary and much researched. The ways in which temperatures are achieved and managed in thermal systems are also the subject of much innovation. Innovations such as these seek to improve the efficiency of production processes. Other innovations, however, are directed at lessening adverse environmental impacts. In Chap. 5, we will consider ways in which to reduce the emissions of harmful greenhouse gases.

In Chap. 1, we explored the role of hydrogen innovation in the petroleum industry, for example, associated with desulphurisation. SO_2 is a noxious gas associated with

Table 3.1 Thermal processes for hydrogen production from fossil fuels

Process	Process variant	Attributes
Steam reforming		Dominant process for hydrogen production—typically from natural gas. The process is endothermic
	Autothermal reforming	Incorporates catalytic partial reforming for heat generation
Partial oxidation (POX)		Exothermic production of CO and hydrogen from any fossil fuel
	Coal Gasification	Particular application of POX to produce syngas for combustion. Hydrogen separation possible, e.g., by use of membranes or adsorbents
Plasma processing		A high-technology approach, involving microwave energy input, producing hydrogen and perhaps potentially valuable carbon nanoforms

the formation of acid rain. In most developed markets, it is mandated that sulphur must be removed from fuels available for retail sale. In addition, many sources of crude oil available to the international oil companies are relatively high which is sulphur and are said to be "sour". Hydrogen is used during refining, for cleaning sour crude oil, via the formation of hydrogen sulphide. It would be innovative, and potentially resource efficient, if another route were available for the desulphurisation of sour crude oil. Research has been devoted to the possibility of very high-temperature supercritical water-based approaches to desulphurisation, known as "hydrodesulphurisation". Supercritical water is so hot that the distinction between water (liquid) and steam (vapour) has been entirely lost. This is another example of potentially beneficial innovation associated with petroleum chemical engineering. Historically, the energy crises of the mid 1970s and the early 1980s motivated much research to convert coal and sour crude to useful fuels using supercritical water (SCW), as an alternative desulphurisation method [7]. More recently, such supercritical water techniques have been generating interest for the upgrading of heavy oil fractions and bitumen. The ability to remove sulphur and other impurities from heavy hydrocarbons is a key feature of SCW upgrading (SCWU) processes. Unlike other desulphurisation methods, SCWU inhibits coke formation, thereby maximising carbon yield [8].

Oxidative desulphurisation (ODS) is another hydrogen-free desulphurisation process. Despite its relative novelty, it has already received substantial interest due to its potential to be an energy-efficient alternative, or complement, to hydrodesulphurisation (HDS) for the removal of sulphur from sour crude oil [9]. ODS comprises two major steps: first chemical oxidation into organic sulphones and second adsorption, or extraction, of the sulphone. The advantages of ODS over HDS are threefold: first it operates under lower temperature and pressure (<100 °C, <100 bar) compared with HDS (>200 °C, >100 bar). Second, no hydrogen is needed in the process, consequently reducing costs of operation, safety risks and other problematic issues associated with the use of hydrogen, and thirdly: it is more effective in removing contaminants (notably thiophenic content) from hydrocarbon compounds when compared with the HDS alternative [9]. However, to achieve deep desulphurisation by these means, while maintaining high fuel recovery, requires the use of an expensive and/or potentially toxic catalyst [10].

According to Ates et al. [8], SCWU has the most potential as a pre-treatment technology to convert heavy, high-sulphur content oils into lighter, low-sulphur feeds prior to further refining. Modern fuel requirements stipulate fuel sulphur levels on the order of 10 ppmw. SCWU by itself is not likely to achieve such low sulphur levels in refined fuel. Therefore, the future of SCWU is likely as a sub-unit within the refinery [11]. In this context, while ODS can complement SCWU by being able to capture the thiophenic compounds that evade SCWU alone, SCWU can complement ODS by acting as an oxidant and hydrogen-free pre-treatment for crude oil prior to ODS, thereby greatly reducing hydrogen requirements. As such, the integration of ODS with SCWU can represent an efficient and economically attractive option [6, 9]. Generally, sulphur removal technologies should be evaluated for effectiveness, sulphur discrimination, economic viability, size, energy efficiency and potential environmental impacts.

It should be remembered that generally petrochemical refinery infrastructures have been highly profitable using well-established hydrogen-based technologies. As such, a major hindrance to innovative hydrogen-free desulphurisation technologies is a perception in the industry that capital investments in hydrogen-free desulphurisation technologies would not be economically viable at this time. It is indeed the case that installing new refinery equipment to meet tighter environmental standards will be capital intensive. Nevertheless, one can see the prospect that existing infrastructures will be reconfigured, for example favouring the use of SCWU as on-site pre-treatment as part of a move to hydrogen-free desulphurisation technology. How these innovative developments might affect the growth of a hydrogen economy is a complex question. If the refining industry turns away from hydrogen will that weaken the industrial underpinnings of the nascent hydrogen economy, or will it liberate supplies to meet growing demand? Might capital investment in new and more environmentally benign refinery operations be part of a more general process in which hydrogen production is cleaned-up? A big shift towards hydrogen production with carbon capture, storage and utilisation would represent a major step by the petroleum industry towards environmental impact reduction, such issues will be considered further in Chap. 10.

We close this chapter by noting that mature methods for hydrogen production rely on advanced and improving chemical engineering techniques involving proprietary catalysts. We suggest that there is ample opportunity for further innovation in that area. Various interesting avenues of research were reported at the WHEC 2016 conference in Zaragoza, Spain. These include sorption-enhanced reforming (SER) aiming for hydrogen production with integrated CO_2 capture in one step. The process involves reformation, the water-gas shift reaction and carbonation [7].

$$CaO(s) + CO_2(g) \rightarrow CaCO_3(s) \tag{3.4}$$

One candidate supporter for nickel catalysts is mayenite ($Ca_{12}Al_{14}O_{33}$) [7]. Aranda et al. have reported on work using CaO-based sorbent for CO_2 capture [8]. They observe:

Steam reforming ΔH 224.8 kJ/Mol
Water-gas shift ΔH -35.6 kJ/Mol
CO_2 capture ΔH -170.5 kJ/Mol.

Aranda et al's method involves CaO finely dispersed on a mayenite support. It is a wet micropowder synthesis method involving thioglycolic acid (TGA). The researchers claim that their wider EU ASCENT project (Advanced Solid Cycles) has looked at the engineering economics of TGA-SER approaches. Broadly, the methods are competitive with Pressure Swing Absorption processing of product gases. They report that they have achieved hydrogen purities of 98–99%.

Fernandez et al. have researched calcium looping and chemical looping combustion showing higher efficiencies and lower equipment costs in SER. They regenerate the sorbent on a 15-min cycle [9].

In summary, there is ample evidence already that the Mature Hydrogen sector is innovative and responsive to incentives. One final example has been the redesign of nozzles for natural gas combustion in SMR boilers, so as to reduce nitrous oxide emissions in the face of tightening air quality regulation.

3.4   Methane and the Atmosphere

Elsewhere in this book (e.g. Sect. 7.1) we will point to futures based on large-scale electrification of the energy system. A key challenge facing such futures is the roll-out of economic large-scale electricity storage technologies, possibly involving hydrogen. The Mature Hydrogen approach based on natural gas also faces key, but different, technical challenges. Prominent among them is the need for the natural gas industry to reduce significantly losses due to gas flaring, venting and leakage (the latter is sometimes termed "fugitive emissions"). Such wastage is particularly problematic as the International Panel on Climate Change has estimated that over short timescales (20 years), methane is 86 times more damaging (per kg) in the atmosphere than carbon dioxide. It has been estimated that these losses more than double the climate impact of the natural gas industry [12]. The more progressive players in the industry recognise the need to act on this problem and have formed the Oil and Gas Climate Initiative to limit methane emissions to 0.25% of total marketed product by 2025. If Mature Hydrogen approaches are to have a chance of making a positive contribution to a low carbon economy, such leaks need to be quite literally tightened up.

References

1. Zou, C., Q. Zhao, and X. Bo. 2016. Energy revolution: From a fossil energy era to a new energy era. *Natural Gas Industry B* 3 (1): 1–11. https://doi.org/10.1016/j.ngib.2016.02.001.

2. EAJV. 2018. *Hydrogen production and purification*. The future of Natural Gas 2018. Available from: http://www.eajv.ca/english/h2.

3. Muradov, N. 2017. Low to near-zero CO2 production of hydrogen from fossil fuels: Status and perspectives. *International Journal of Hydrogen Energy* 42 (20): 14058–14088.

4. Van Hook, James P. 1980. Methane-steam reforming. *Catalysis Reviews* 21 (1): 1–51. https://doi.org/10.1080/03602458008068059.

5. Bakenne, A., W. Nuttall, and N. Kazantzis. 2016. Sankey-diagram-based insights into the hydrogen economy of today. *International Journal of Hydrogen Energy* 41 (19): 7744–7753.

6. McWilliams, A. 2018. *The global hydrogen economy: technologies and opportunities through 2022*. BCC Research.

7. Kida, Y., et al. 2014. Combining experiment and theory to elucidate the role of supercritical water in sulfide decomposition. *Physical Chemistry Chemical Physics* 16 (20): 9220–9228.

8. Ates, A., et al. 2014. The role of catalyst in supercritical water desulfurization. *Applied Catalysis, B: Environmental* 147: 144–155.

9. Timko, M.T., et al. 2014. Response of different types of sulfur compounds to oxidative desulfurization of jet fuel. *Energy & Fuels* 28 (5): 2977–2983.

10. Gonzalez, L.A., et al. 2012. oxidative desulfurization of middle-distillate fuels using activated carbon and power ultrasound. *Energy & Fuels* 26 (8): 5164–5176.

11. Timko, M.T., A.F. Ghoniem, and W.H. Green. 2015. Upgrading and desulfurization of heavy oils by supercritical water. *The Journal of Supercritical Fluids* 96: 114–123.

12. Davis, M. 2019. Monetising flared gas … innovative applications of proven technology, presentation to EPRG Summer Conference, London, 2019. Slides available at: https://www.eprg.group.cam.ac.uk/wp-content/uploads/2019/09/M.-Davis_2019_upd.pdf.

Chapter 4
Towards a Hydrogen Economy

This chapter is based upon unpublished work by Bernardo Castro-Dominguez (University of Bath, UK) and Nikolaos Kazantzis (Worcester Polytechnic Institute, USA). We are most grateful to them for making their work available to us.

4.1 Origins

In 1923, the British scientist and polymath J. B. S. Haldane imagined the years to come. He addressed the Cambridge Heretics Society and presciently noted [1].

> Personally, I think that four hundred years hence the power question in England may be solved somewhat as follows: The country will be covered with rows of metallic windmills working electric motors which in their turn supply current at a very high voltage to great electric mains. At suitable distances, there will be great power stations where during windy weather the surplus power will be used for the electrolytic decomposition of water into oxygen and hydrogen. These gasses will be liquefied, and stored in vast vacuum jacketed reservoirs, probably sunk in the ground. If these reservoirs are sufficiently large, the loss of liquid due to leakage inwards of heat will not be great; thus the proportion evaporating daily from a reservoir 100 yards square by 60 feet deep would not be 1/1000 of that lost from a tank measuring two feet each way. In times of calm, the gasses will be recombined in explosion motors working dynamos which produce electrical energy once more, or more probably in oxidation cells. Liquid hydrogen is weight for weight the most efficient known method of storing energy, as it gives about three times as much heat per pound as petrol. On the other hand it is very light, and bulk for bulk has only one third of the efficiency of petrol. This will not, however, detract from its use in aeroplanes, where weight is more important than bulk. These huge reservoirs of liquified gasses will enable wind energy to be stored, so that it can be expended for industry, transportation, heating and lighting, as desired. The initial costs will be very considerable, but the running expenses less than those of our present system. Among its more obvious advantages will be the fact that energy will be as cheap in one part of the country as another, so that industry will be greatly decentralized; and that no smoke or ash will be produced.

© Springer Nature Switzerland AG 2020
W. J. Nuttall and A. T. Bakenne, *Fossil Fuel Hydrogen*,
https://doi.org/10.1007/978-3-030-30908-4_4

However, as noted in Chap. 1, it was not until the early 1970s that the term "hydrogen economy" entered the lexicon predominantly as a means to describe the use of hydrogen as a clean fuel for vehicles. The idea gained some traction during the oil shock of 1973–1974 as a way to help free the west from economic dependence on, and political linkage to, the Middle East. As things turned out, however, the hydrogen economy did not emerge in those years beyond the expanding use of hydrogen in the petroleum and chemical sectors, as described in earlier chapters. Rather than a transition to hoped-for era of clean and secure energy, we have instead seen fifty years of continuing strife and concern relating, in particular, to access to Persian Gulf oil resources.

Over the years, considerable attention has been given to the economic performance of various production options. As noted in Chap. 1, in 2017 a group of companies formed the Hydrogen Council in order to promote hydrogen as a key element is fostering the global energy system's transition to a sustainable, robust and economically viable future state [2]. Within such a context, various economic performance studies on the production, storage and infrastructure of hydrogen have been presented. While the concerns of the 1970s did not generate the hoped-for emergence of the new hydrogen economy, the growing concerns around environmental security, urban air quality and, above all, climate change might be sufficient for the shift to finally occur. Hydrogen production is increasingly being recognised as a key enabler of a cleaner and more secure energy supply. This has provided ample motivation for the pursuit of various comprehensive economic performance assessment studies for different incumbent hydrogen technology options as well as of emerging technologies currently at the demonstration stage. As noted in Chap. 1, the sources of hydrogen production can be classified as "mature" or fossil fuel-based (natural gas, coal) and "green" or renewable (e.g. wind, sunlight, biomass) [3]. There is also a potential for nuclear energy to play a role in hydrogen production [3]. As things stand, however, nuclear power for electricity is struggling to find an economically viable role in electricity markets.

4.2 Emergence

Figure 4.1 illustrates a low-carbon scenario from Shell consistent with the Paris Agreement of the United Nations Framework Convention on Climate Change. The Shell SKY Scenario shows hydrogen emerging as a major energy carrier for industry and transport after 2040. In the same scenario, natural gas plays the role of a transition fuel showing a decline in natural gas use in many markets including Europe and North America out to 2050 (see Chap. 3).

Andrew McWilliams, writing for BCC Research, has observed that global hydrogen economy investments in 2016 exceeded $5.1 billion. He predicts that this will rise to $14.1 billion by 2022, representing a compound annual growth rate of 18.6% in this period [5].

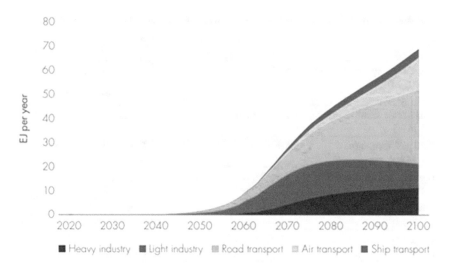

Fig. 4.1 Global hydrogen demand in the low-carbon Shell Scenario: *SKY—meeting the goals of the Paris Agreement* [4]

In Chap. 1, we introduced the notion that if one were to compare Green Hydrogen with Mature Hydrogen and if we require minimal greenhouse gas emissions from both options, i.e. to require the use of carbon capture and storage (CCS) for Mature Hydrogen, then one sees that the cost ratio of the two approaches is roughly 1.8, in favour of the Mature Hydrogen approach. That assessment was derived from our earlier published work [6]

Other workers have also looked at such issues including Paul Ekins and colleagues in 2010, as shown in Table 4.1. Their work highlights the potential cost competitiveness of biomass as a hydrogen source. While we see clear merit in such an approach, we wonder whether biomass-based methods could ever meet the scale of the challenge. With that said, we note the impressive progress made by blended biofuels in today's liquid fuels' supply.

4.3 Castro-Dominguez/Kazantzis Review 2018

In this section, we consider recent work by Bernardo Castro-Dominguez and Nikolaos Kazantzis. Their review of hydrogen economics is summarised in Table 4.2. In order to understand the various methods assessed, it is useful to recap over the key hydrogen production methods, some of which have been described in more detail in earlier chapters.

Steam reforming of natural gas, also known as steam methane reforming (SMR), is a technologically mature and well-established method for generating hydrogen. The efficiency of the process has been estimated to range between 60

Table 4.1 Cost of hydrogen production based on work of Ekins et al. [7]

Technology	Cost range US$ (year 2000)/GJ of hydrogen	Note
Large-scale steam reforming (>1000 MW)	5.25–7.26	For steam reforming natural gas price is a key consideration
Small-scale steam reforming (<5 MW)	11.50–40.40	Smaller scale, higher cost
Coal gasification (min. 376 MW)	5.40–6.80	Carbon dioxide capture would add 11% to costs
Biomass gasification (>10 MW)	7.54–32.61	Average cost 14.31
Biomass pyrolysis (>10 MW)	6.19–14.98	Cost reduced by sale of co-products
Large-scale electrolysis (>1 MW)	11.00–75.00	Cost of electrolysis is a key consideration
Small-scale electrolysis (<1 MW)	28.00–133.00	Cost is highly size dependent—smaller scale, higher cost

and 80% lower heating value (LHV) [8], with capacities of 5000 and 250,000 m^3 STP/h [3, 9, 10]. SMR is indeed a flexible technology option that enables process operation at different capacities; this characteristic allows the development of small-scale reformers, which are considered suitable for refuelling stations and thus relevant to develop an efficient hydrogen economy [8, 11]; see also Chap. 8.

Large-scale steam reforming plants without carbon capture have an estimated levelised production cost ranging between $1.30 and $3.17 per kg of H_2 [10, 12–14] (in Chap. 3, we noted a possible cost of $1.30 per kg of H_2).

Hydrogen production costs via steam reforming vary according to location and time; therefore, many economic performance assessment studies have utilised standard engineering methodologies to adjust their estimations such as those presented by the U.S. Department of Energy [14], the International Energy Agency (IEA) [15] and others [13]. The latest revised economic performance assessment studies include the work of Salkuyeh (2017) [16] whose costs were estimated to be between $1.30 and $1.50 per kg of H_2 and a technical report (IEAGHG) [10] from the IEA (2017) where costs were estimated to be $1.40 per kg of H_2.

The incorporation of carbon capture and sequestration (CCS) systems to mitigate greenhouse gas emissions has been considered by several economic performance evaluations. The implementation of CCS systems has a significant impact on economic performance and characteristics. Salkuyeh [10] has concluded that the inclusion of carbon capture units decreases the efficiency by 18%, and it has been estimated that hydrogen production costs rise (if one accepts a $1.3 starting point) to $2.10–$2.27 per kg of H_2 [12, 13]. The cost estimated by the IEA (2017) ranges between $1.68 and $2.06 per kg of H_2, with key increments in operating costs ranging between 18 and 33% [17].

Table 4.2 Economic performance characteristics found in the pertinent literature for various hydrogen production systems, as collated by Castro-Dominguez and Kazantzis and previously unpublished

Technology	Efficiency (%)	Costs [$/kg of H_2]	References
Natural gas reforming w/o CCS	72	1.30	[13]
	74–85	2.08	[12]
		1.84	[3, 18]
		3.01	[3, 25]
		2.08	[3, 20]
		1.40	[17]
Natural gas reforming w CCS	71	2.10	[13]
		2.27	[3, 12]
		1.68–2.06	[17]
Coal gasification w/o CCS	56	1.30–1.70	[13]
		1.34	[12]
		0.37–1.25	[3, 18]
		0.78	[3, 18]
		1.34	[3, 20]
Coal gasification w CCS		1.80–2.40	[13]
		1.63	[3, 20]
		1.02	[3, 19]
Biomass gasification	48	2.10–2.30	[13]
		1.77–2.05	[3, 12, 23]
		1.06–1.86	[3, 22]
		1.59–1.70	[12]
Biomass pyrolysis	35–50	1.25–2.20	[12]
Water electrolysis alkaline	62–82	4.10–5.50	[13]
Water electrolysis (PEM)	65–78	4.10–5.50	[13]
Water electrolysis (SOEC)	80–90	2.80–5.80	[13]
Water electrolysis (Solar)	40–60	5.78–23.2	[12]
Water electrolysis (nuclear)	40–60	4.15–7.00	[3, 12, 26]

Partial oxidation coal gasification (POX), as outlined in Chap. 3, is another fossil fuel-based process that is commonly used for the generation of hydrogen. In this process, coal is thermally treated at high temperatures in the presence of oxygen. Coal gasification is a mature process, but less used than the SMR approach. The efficiency of the POX process ranges between 54 and 75% (LHV) [3, 9, 13, 18] depending on the design and capacities of 5000 and 20,000–100,000 m^3STP/h [13]. The estimated levelised production costs for gasification plants that utilise coal without carbon caption are: $0.83–$1.7 per kg of H_2. [3, 18]. Certainly, coal is a cheaper fuel than natural gas, but the POX method requires higher capital investments than SMR and hence only large-scale plants are considered [8]. Other considerations are also described in Chaps. 1, 5 and 6.

Similar to plants utilising SMR, there has been a lot of interest in coupling coal gasification with CCS systems. For instance, in 1998, Gray and Tomilson have presented costs of $0.93 and $0.78 per kg of H_2 with, and without, carbon capture and sequestration systems. Similarly, in 2002, Kreutz et al. [19] estimated respective costs of $1.02 (with CCS) and $0.86 (no CCS) per kg of H_2; while in 2005, Rutkowski [20] estimated the respective costs at $1.63 and $1.34 per kg.

Biomass gasification converts organic materials such as plants, wood, and/or waste, into syngas (H_2 and CO) via thermochemical treatment in the presence of oxygen. This process has gained much attention over the last 25 years [21]. The efficiency of this process has been estimated to be between 47 and 48% (LHV) [13] while operating at medium-size capacities of 57,000 m^3STP/h [9]. Biomass gasification produces H_2 at costs estimated to range between $1.44 and $2.83 per kg of hydrogen [8, 9, 12, 13, 22]. The cost depends significantly on the type of feedstock used. This renewable technology option remains somewhat more expensive than fossil fuel-based H_2 production methods such as coal gasification and natural gas steam reforming, but could be boosted significantly by relatively modest policy support.

The related, but more sophisticated process of biomass pyrolysis converts organic materials into char, oil and gases through chemical treatment in the absence of oxygen. Studies, such as the one presented by Padro and Putsche [22], have estimated that biomass pyrolysis has an estimated production cost of $2.57 to $1.25 per kg of H_2, which is slightly better than is achieved by biomass gasification.

Water electrolysis involves the dissociation of the water molecule to generate H_2 and O_2. The most common technologies used for electrolysis rely on alkaline, proton-exchange membranes (PEM) and solid oxide electrolysis cells (SOEC). The efficiency of water electrolysis is strongly correlated to the type of electrolyser used. For instance, the efficiency of alkaline, PEMs and SOECs ranges between 65–82%, 65–78% and 80–90%, respectively. The amount of electricity required by the electrolysers is high, thus making this technology difficult to compete with well-established technologies such as steam reforming of natural gas.

Ogden [13] has utilised the DOE H2A Production Analysis [23] approach to estimate the costs for producing H_2 via three different types of electrolyser. The costs ranged between $4.1 and $5.5 per kg of H_2 for both alkaline and PEMs and between $2.8 and $5.8 per kg of H_2 for SOECs.

The economic features of water electrolysis vary significantly when the electricity needed to dissociate water is supplied by alternative energy sources (such as hydro, wind, solar and nuclear) or a combination of them. For instance, the most expensive source is solar PV, which can produce H_2 with a maximum capacity of 360,000 m^3STP/h at \$10.49 per kg of H_2. The cheapest source is through the use of nuclear energy; it has a capacity of 265,000 m^3STP/h and a production cost of \$4.15 per kg of H_2 [12]. Bockris and Veziroglu [24] reported that electrolysis can produce H_2 at price of \$3.50 per kg of H_2 utilising wind; nonetheless, the capacity of such a plant was not mentioned.

Advocates of Green Hydrogen sometimes point to the in-principle possibility of using otherwise surplus (and hence with a very low market price) renewable electricity at times of high production (i.e. suitable weather) and low demand. The appropriateness of using such costs rather than total system costs can be debated, especially noting that in some territories, such as the UK all renewable power is purchased at fixed (Contract for Difference) prices. Such calculations and the under-pinning assumptions can be rather complicated. Additionally, in such thinking if cheaply priced renewable electricity can be obtained, then attention must be given to the proportion of the time that advantageously cheap electricity is actually likely to be available. Furthermore, attention should also be given to the electricity ramp-up rates required by electrolysers, especially for steam electrolysis and this compared to the duration of low electricity prices. It seems to us that it is probably inappropri-ate ever to regard renewable electricity as free, or almost free at a social or policy level. Additionally, whether an investor might be able to navigate subsidies and mar-ket design to find cheap prices is a different question. Generally, we note that the substantial underlying fixed costs of large-scale renewable electricity systems have been met by some form of energy policy subsidy, cost socialisation or risk guarantee. These market adjustments can make Green innovations attractive opportunities for investors, but economically they can sometimes be less attractive when considered in terms of the wider social costs.

We note that in principle hydrogen might be produced in future via catalytic thermal approaches. Candidate technologies include the sulphur–iodine cycle and the calcium–bromine cycle. These approaches have been tested under laboratory conditions, but thus far have not been commercialised at an industrial scale. For this reason, and the associated truth that costs are very hard to estimate, these approaches are excluded from consideration in Table 4.2.

4.4 Looking to the Future

From the analysis above, we see that SOEC-generated Green Hydrogen has an aver-age cost of \$4.3 per kg to be compared with roughly \$2.0 per kg for Mature Hydrogen (via SMR and with CCS—see Table 4.2). From these figures, we infer a cost ratio of roughly 2.1 in favour of mature hydrogen compared to the cost ratio of 1.8 that we introduced in Chapter 1. It remains clear that SMR hydrogen from natural gas with CCS is for now a more attractive proposition than Green Hydrogen.

There is a widespread sense, akin to the notion of natural gas is a transition fuel, that the journey towards a Green Hydrogen future might be accompanied by a temporary and transitional use of Mature Hydrogen as the Green Hydrogen supply chain develops.

One driver of a greener future has been the use of rising "carbon prices" associated with greenhouse gas emissions. The implication is that once a very high carbon price is established, then SMR-based Mature Hydrogen will be replaced by cleaner Green Hydrogen. Such a conclusion deserves closer scrutiny.

First let us examine a world with established Mature Hydrogen facing a shift to cleaner Green Hydrogen. Let us assume the cost of renewably generated Green Hydrogen to be \$4.0/kg H_2 and the cost of unabated (no CCS) Mature Hydrogen to be \$1.3/kg H_2 (see observations earlier in this chapter and [9]), then the cost difference between the two processes is US \$2.7/kg H_2. The unabated SMR process yields 7.33 kg of CO_2 for each kg of H_2 produced [6] [and references therein]. This implies that the emission of one tonne of CO_2 is associated with the production of 150 kg of unabated Mature Hydrogen. The cost difference between the Green and Mature alternatives for 150 kg of H_2 is \$405. So up to a GHG emissions price of \$405/Tonne of CO_2 the Mature Hydrogen alternative is economically more attractive than the Green Hydrogen alternative. A carbon price of \$400/Tonne of CO_2 remains far above anything seen worldwide; indeed, the EU-Emissions Trading Scheme price has rarely been above €20/Tonne of CO_2 (\$23/Tonne of CO_2).

Now let us consider the carbon price incentive necessary to motivate a shift from unabated Mature Hydrogen towards bringing in the use of CCS technology. If CCS reduces the emissions by 85%, then the emitted CO_2 per tonne of hydrogen-produced drops from 7.33 tonnes to 1.1 tonnes. The emission difference between the unabated and abated approaches is, therefore, 6.2 tonnes of CO_2 per tonne of H_2. One tonne of abatable CO_2 relates to the production of 161 kg of SMR H_2. Assuming the cost of abated (SMR+CCS) H_2 to be \$2.0/kg H_2, the cost difference with respect to unabated H_2 is \$0.70/kg H_2. The production of one tonne of potentially abatable CO_2 corresponds to an abatement cost of \$113. That is with carbon prices above roughly \$113, one can expect SMR hydrogen producers to consider shifting to CCS abatement. This is a much more economically and politically plausible step than the Green Hydrogen shift considered earlier.

If technology costs do not change, then as carbon prices rise the incentive will be to adopt CCS approaches rather than to shift to Green Hydrogen approach. Public policies and subsidies would need to be substantial to overcome these economic realities if renewables and electrification are to win through, and to what environmental benefit? In 2019 there remains much uncertainty about the prospects for both Green Hydrogen and the abated Mature Hydrogen (SMR+CCS) approaches. Challenges on both sides include feasibility at scale. The Mature Hydrogen approach would appear to come at a lower cost but will it represent a sufficiently low emission option for it to be regarded as firmly part of the solution in a low carbon future? Key to this aspect will be success with CCUS (see Chap. 5). It should further be stressed that across the full life-cycle neither approach is zero-CO_2, both are merely low-CO_2.

In critiquing the Green Hydrogen approach, we have assigned no value to the oxygen that can be obtained from water electrolysis, and indeed, it is usually vented. If a value could be achieved for this product, then the economics of the Green Hydrogen approach become somewhat more favoured.

Another point to concede is the high levels of uncertainty in the input numbers used in the estimations above. With different numbers, different conclusions would emerge.

In his 2017 review (see Chap. 3), Nazim Muradov considers the costs of CO_2 abatement from two possible technologies: one he describes as a "new hydrogen plant" and the other is a natural gas-fuelled combined cycle gas turbine (CCGT) [27]. Both technologies are envisaged to operate with carbon capture and storage (CCS). He observes that fitting CCS to the CCGT plant increases product costs by 46%, whereas for the new hydrogen plant, the equivalent figure is only 15%. The cost of CO_2 avoided from the CCGT plant with CCS is US$ 53 per tonne of CO_2, to be compared with US$ 15 for the new hydrogen plant. Much of the benefit of the hydrogen-based approach relates to the richer CO_2 concentrations encountered in hydrogen SMR systems (see Chap. 3). In this way, we see that the cleaning up of a hydrogen plant is a far more attractive proposition than cleaning up the natural gas-fuelled fossil fuel alternative.

In this way, we hope to have revealed the low-carbon economic case for hydrogen from natural gas and most especially the affordable environmental benefits that can flow from such decisions.

References

1. Shalizi, C.R. 1993. *DAEDALUS or science and the future*. C. A paper read to the Heretics, on February 4, 1923, Berkeley, California.
2. Hydrogen Council. 2018. Available from: http://hydrogencouncil.com.
3. Bartels, J.R., M.B. Pate, and N.K. Olson. 2010. An economic survey of hydrogen production from conventional and alternative energy sources. *International Journal of Hydrogen Energy* 35 (16): 8371–8384.
4. Shell Scenarios SKY Meeting the Goal of the Paris Agreement. January 2, 2019. Available from: https://www.shell.com/energy-and-innovation/the-energy-future/scenarios/shell-scenario-sky.html.
5. McWilliams, A. *The global hydrogen economy: Technologies and opportunities through 2022*. 2018, BCC Research.
6. Bakenne, A., W. Nuttall, and N. Kazantzis. 2016. Sankey-Diagram-based insights into the hydrogen economy of today. *International Journal of Hydrogen Energy* 41 (19): 7744–7753.
7. Ekins, P., S. Hawkins, and N. Hughes. 2010. Hydrogen technologies and costs. In *Hydrogen energy—Economic and social challenges*, 29–57.
8. Dodds, P.E. and W. McDowall. 2012. *A review of hydrogen production technologies for energy system models*. UCL Energy Institute. Available from: http://www.wholesem.ac.uk/bartlett/energy/research/themes/energy-systems/hydrogen/WP6_Dodds_Production.pdf.

9. Mueller-Langer, F., et al. 2007. Techno-economic assessment of hydrogen production processes for the hydrogen economy for the short and medium term. *International Journal of Hydrogen Energy* 32 (16): 3797–3810.
10. Khojasteh Salkuyeh, Y., B.A. Saville, and H.L. MacLean. 2017. Techno-economic analysis and life cycle assessment of hydrogen production from natural gas using current and emerging technologies. *International Journal of Hydrogen Energy* 42 (30): 18894–18909.
11. Ogden, J.M. 2001. *Review of small stationary reformers for hydrogen production.* New Jersey: Center for Energy and Environmental Studies.
12. Nikolaidis, P., and A. Poullikkas. 2017. A comparative overview of hydrogen production processes. *Renewable and Sustainable Energy Reviews* 67: 597–611.
13. Ogden, J.M. 2018. *Prospects for hydrogen in the future energy system.* California: Institute of Transportation Studies.
14. Ruth, M. and F. Joseck. 2011. *Hydrogen threshold cost calculation,* ed. S. Satyapal. Department of Energy, U.S.A.
15. Williams, R.B., et al., *Estimates of hydrogen production potential and costs from California Landfill Gas., May 7–11, 2007.* In *Proceedings of 15th European Biomass Conference and Exhibition,* Berlin, Germany.
16. Technology Roadmap Hydrogen and Fuel Cells. 2015. Available from: www.iea.org/publications/freepublications/publication/TechnologyRoadmapHydrogenandFuelCells.pdf.
17. Collodi, G., G. Azzaro, and N. Ferrari. 2017. Techno-Economic Evaluation of SMR Based Standalone (Merchant) Hydrogen Plant with CCS. In *IEAGHG Technical Report,* Cheltenham, UK.
18. Gray, D. and G. Tomlinson. 2001. *MTR 2002-31 Mitretek Technical Paper HYDROGEN FROM COAL.*
19. Kreutz, T., et al. 2005. Co-production of hydrogen, electricity and CO2 from coal with commercially ready technology. *Part B: Economic Analysis* 30: 769–784.
20. Rutkowski, M. 2005. *Hydrogen from coal without CO_2 capture and sequestration.* Available from: www.hydrogen.energy.gov/h2a_prod_studies.html.
21. Molino, A., S. Chianese, and D. Musmarra. 2015. *Biomass gasification technology: The state of the art overview,* vol. 25.
22. Padro, C.E.G. and V. Putsche. 1999. *Survey of the economics of hydrogen technologies,* Colorado.
23. DOE H2A Production Analysis. Hydrogen and Fuel Cells Program 2018. Available from: https://www.hydrogen.energy.gov/h2a_production.html.
24. Bockris, J.O.M., and T.N. Veziroglu. 2007. Estimates of the price of hydrogen as a medium for wind and solar sources. *International Journal of Hydrogen Energy* 32 (12): 1605–1610.
25. Penner, S.S. 2006. Steps toward the hydrogen economy. *Energy* 31 (1): 33–43.
26. C. Petri, M., B. Yildiz, and A. E. Klickman. 2006. *US work on technical and economic aspects of electrolytic, thermochemical, and hybrid processes for hydrogen production at temperatures below 550 °C,* vol. 1.
27. Muradov, N. 2017. Low to near-zero CO2 production of hydrogen from fossil fuels: Status and perspectives. *International Journal of Hydrogen Energy* 42 (20): 14058–14088.

Chapter 5
Carbon Capture, Utilisation and Storage

The EU emissions trading system (EU-ETS) is a major part of the European Union's policy to combat climate change. It covers approximately 45% of the EU's total GHG emissions (more than 11000 power station and industrial plants in 31 countries, as well as airlines) [1]. In 2014, emissions covered under the EU-ETS amounted to 1 868 Mt CO_2–eq. Power plants and other industrial installations covered by the EU-ETS jointly emitted 97% of the total, 1813 Mt CO_2–eq. in 2014, and the remainder being attributed to aviation activities (3%, 55 Mt CO_2–eq.) [2]. As shown in Fig. 5.1, the main source of emissions in the EU-ETS is the combustion of fuel, occurring mainly in power and heat plants. Combustion installations emitted 67% (1218 Mt CO_2–eq.) in 2014, while the refinery and chemical industry emitted a 15% share (453 Mt CO_2–eq.) of the total verified emissions from stationary installations [2].

The European Union Emissions Trading Scheme has had a difficult history. It has never achieved the carbon price that was expected and intended by policy-makers. In its history, the price has crashed. Once resuscitated, the scheme was still highly volatile and never achieved prices needed to drive beneficial technology-neutral change. The scheme was flawed from its birth—too many permits were granted to historical emitters, a global financial crisis in 2008/2009 led to a significant decrease in energy (and hence carbon) demand and the co-existence of a separate policy mandate to deploy significant amounts of low carbon renewable energy, at any price, further weakened expected demand for permits. The use of a cap-and-trade approach was politically attractive simply because it did not involve the toxic word "tax", but a tax would have been far preferable. First the intended and hoped-for price could have been fixed, and then the emission reductions would have been a beneficial consequence of that decision. However, with the policy, as adopted, it was the total emission reduction that was fixed, which had the unfortunate consequence that what-ever one chose to do within the sector would not reduce emissions, it would simply change the EU-ETS price.

In 2019, there is finally some hope again for the EU-ETS. A "Market Stability Reserve" has been created which has the ability to remove emissions from the cap. In effect, the cap is no longer fixed. The long-term trajectory for emissions now depends on market outcomes. In the short term, there has been a beneficial consequence—by

© Springer Nature Switzerland AG 2020
W. J. Nuttall and A. T. Bakenne, *Fossil Fuel Hydrogen*,
https://doi.org/10.1007/978-3-030-30908-4_5

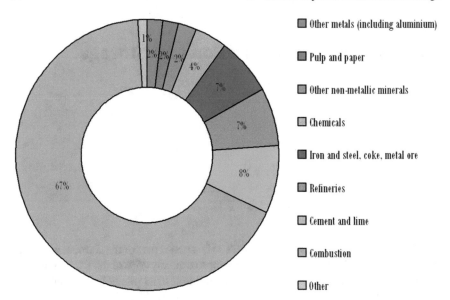

Fig. 5.1 Shares of European Union Emissions Trading System (EU-ETS) emissions by main activity. The largest component corresponds to combustion. Then going clockwise "other" is 1% and other metals 2%. Then, the entries can be read from the key in sequence. *Source* [2]

May 2019, the EU-ETS price had reached Euro 26 per tonne of CO_2, higher than has been seen for more than 10 years and close to historic maxima.

As we have seen, hydrogen from fossil fuel sources already has significant industrial applications in, for example the oil industry and agricultural fertiliser manufacture. If hydrogen is to secure a major role in the wider economy (including personal mobility and domestic heating), then it must be on the basis of viable low-carbon credentials. If the world is to ensure low greenhouse gas emissions from a fossil-fuelled hydrogen economy, then carbon capture and storage will have an important role to play.

5.1 The Case for Carbon Capture and Storage

Generally, CCS is conceived as a means to curb CO_2 emissions from fossil-fuel-based power generation and is the only option available to significantly reduce direct emissions from many industrial processes. Two commonly identified impediments to the widespread deployment of CCS include the cost of implementing CCS and a lack of regulation addressing unique CO_2 storage issues. Carbon prices are affected by legislative measures, e.g. cap-and-trade or a carbon tax. For CCS to be economically viable, a significant CO_2 emission price (such as a strengthened EU-ETS price) would be needed. The European Technology Platform for Zero Emission Fossil Fuel Power

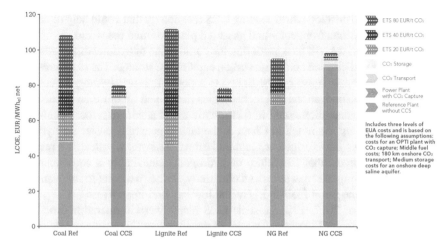

Fig. 5.2 Levelised cost of electricity of integrated CCS power plants compared with unabated alternatives [3]

Plants in 2011 estimated that for power generation from natural gas, the EU-ETS price may need to be as high as €80/tonne CO_2 for CCS to be economically viable in that case (see Fig. 5.2) [3]. The figure of €80/tonne CO_2 compares to a market price in compliance-based markets of around €20 per tonne in Europe [4]. Without sufficient high pricing, CCS will be dependent on subsidies. The profitability of the CCS project is mainly influenced by the CO_2 price, for instance with a distance of 250 km and capacity of $10MtCO_2$/year, the initial CO_2 price has to increase to $46/t for viability [5].

The International Energy Agency (IEA) has declared that "CCS investment needs an urgent boost" identifying "an investment requirement of more than USD 60 billion" [6]. A recent audit by the Global Carbon Capture and Storage Institute found that only 62 of 213 active, or planned, projects were fully integrated commercial-scale projects, of which only seven were actually operating [7]. The scarcity of active projects for CCS is largely attributable to high anticipated costs. While subsidies have been offered to such schemes, they have been revealed to be insufficient to justify progress. For example, in 2010, the Longannet carbon capture project in the UK collapsed. There had been hope that this subsidised project might go ahead, but it was undermined by the imposition in the UK of a carbon price floor for its residual emissions. The carbon price floor tax added around £250 m extra cost to what had previously been a £1bn investment [8]. The UK government was going to partner with Scottish Power, Shell and the National Grid in order to invest the necessary £1bn, but with the new carbon price floor (arising from the Energy Act 2013), it would then take away £250 m as tax. Effectively, the government would have been giving with the one hand and taking away with the other. The cancellation of the project was unfortunate, as it seems likely that this project could have been the best early chance for UK to develop world-leading CCS technology that could help reduce the majority of emissions from coal- and gas-fired plants around the world.

In this chapter, we will be considering the combustion of hydrocarbon fuels, such as natural gas and gasified coal, with carbon capture and storage. The ideas presented may not immediately appear to be linked to a possible future hydrogen economy, but we posit that innovations and technologies developed by those concerned with CCS for power generation could in fact greatly assist a shift to a fossil-fuel-based hydrogen economy. As noted in Chap. 3, the task is generally easier for hydrogen process emissions than for power plant emissions. As such, we present a response to those that suggest that the future of hydrogen is inevitably a consequence of innovations relating to renewables and electricity. Rather we point to innovations at the interface of the fossil fuels industry and low carbon imperatives.

At the close of 2017, 17 large-scale CCS plants were in operation around the world with a further 22 in various stages of development [7]. A large-scale facility is defined as one capable of capturing, transporting and storing at least 800,000 tonnes of CO_2 annually. While this may appear impressive and substantial, Stuart Haszeldine of Edinburgh University has commented that "the development of CCS globally is about 100 times slower than needed to achieve the [Paris Agreement] 2050 climate pledge" [9]. In this book, we posit that if hydrogen from fossil fuels is to have a strong future, it will necessarily be in conjunction with the large-scale expansion of CCS capabilities. We have seen elsewhere in this book the underlying sense of competition between energy futures based upon renewables, electrification and electricity storage and similarly low-carbon futures based upon traditional energy resources, hydrogen as an energy vector and the use of fuel cells for electricity-based mobility. The economic and technical viability of CCS will lie at the heart of that competition.

It is not the purpose of this small book to provide a full overview of issues in carbon capture and storage (CCS). For such insights, we recommend [10, 11]. There are, however, some key considerations associated with conventional approaches to CCS that deserve emphasis, and these are presented in Table 5.1. When considering conventional approaches, we are referring to the use of CCS technology to reduce emissions associated with fossil fuel (natural gas or coal)-fuelled electricity generation.

A key consideration relating to the implementation of power generation with CCS concerns the CO_2 concentration of the exhaust gases emerging from the combustion chamber. The emphasis thus far has been on post-combustion approaches because, simply put, this represents an add-on to conventional power generation systems. For example, a modern combined cycle gas turbine (CCGT) burning natural gas for electricity generation relies throughout on air as the source of oxygen for combustion (Fig. 5.3). Air, however, is only 21% oxygen (by volume). Mostly air consists of nitrogen (78% by volume).

The dry flue gas emitted by a typical CCGT is approximately 11% CO_2 by volume (up from 0.04% in the incoming air) [13]. The key issue is this CO_2 remains diluted in nitrogen. Matching such equipment to a CCS capability requires CO_2 separation from the residual flue gas. The dominant proposal in that regard is termed "carbon

Table 5.1 Considerations associated with carbon capture and storage in power generation

Key consideration	Description
Post-combustion CO_2 capture and storage	It involves the capture of CO_2 from the flue gas stream exiting the combustion process. This mode of capture is applicable for most existing power and chemical plants. Air is used for combustion, the majority of the flue gas is nitrogen, and CO_2 concentrations are low. Thus, the energy for post-combustion capture is the energy required to separate CO_2 from N_2, moisture, and other contaminants in the flue gas
Oxy-fuel combustion	This uses enriched oxygen instead of air for combustion. Thus, one must first enrich oxygen from air. The flue gas from such a process has minimal nitrogen content, so oxy-combustion CO_2 capture involves mere condensation of water from the flue gas. Neglecting the energy for this condensation, oxy-combustion capture energy is just the energy required for the prior air separation
Pre-combustion carbon separation	It involves gasification of a fossil fuel via enriched oxygen to obtain a mixture of CO and H_2. This mixture is converted to a CO_2-H_2 mixture via the water-gas shift reaction. H_2 is separated from the CO_2 and is combusted to generate heat or power. In contrast to the first two modes, pre-combustion capture involves two separations, but with lower energy requirements. The first is to enrich oxygen from air, and the second is to separate CO_2 from H_2. Thus, pre-combustion capture energy is the sum of the energies for air and CO_2-H_2 separations
Storage and sequestration	Storage and sequestration refer to the process of long-term storage of CO_2 after it has been captured and concentrated from various sources. In geological sequestration, the captured and concentrated CO_2 stream is compressed, transported to and then stored in suitable geological sites. The potential geological sites used for the long-term storage of CO_2 include depleted oil and gas reservoirs, spent coal seams, deep saline formations and oceans. In comparison to utilisation, CCS is usually more economical, and there have been more research and demonstration sites dedicated to it. Lack of storage sites close to the emission sources and the high costs associated with the long-distance transport of CO_2 are some of the factors currently limiting the deployment of CCS. It must be conceded that while there are numerous CCS projects around the world there remains some scientific uncertainty around the geochemistry and the practicalities of robust and secure long-term storage of carbon dioxide. Such issues will be of profound importance for the long-term viability of abated Mature Hydrogen strategies (SMR+CCS)

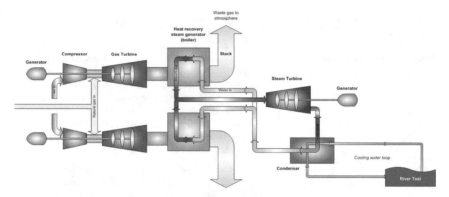

Fig. 5.3 Schematic of combined cycle gas turbine power plant. *Source* Marchwood Power—with permission [12]

scrubbing"—the use of amines to extract CO_2 by adsorption [13]. As Gary Rochelle noted in 2009: "The minimum work requirement to separate CO_2 from coal-fired flue gas and compress CO_2 to 150 bar is 0.11 megawatt-hours per metric ton of CO_2" [14]. Rochelle further notes that in 2006 such approaches might be expected to separate 74% of flue gas CO_2. The power generation penalty and the relatively low efficiency of the process have militated against its effectiveness in moves towards a low carbon economy, especially when one confronts the significant progress made by low emission renewable approaches to power generation in recent years albeit based in-part upon substantial electricity sector cost socialisation.

It is important to stress that any economic and technical concerns arising from the relatively low-concentration of CO_2 in fossil fuel power plant flue gas would not apply so severely to the process emissions from an SMR or POX hydrogen production facility. As discussed previously, these emissions are much richer in CO_2 greatly favouring the use of CCS and CCUS approaches.

In principle pre-combustion, CO_2 is more efficient (in technical terms at least). The engineering challenges are, however, substantial. The key to pre-combustion approaches is to separate the useful oxygen from the predominant nitrogen in the air. Once separated the fuel, for instance natural gas (methane) can, theoretically at least, be combusted in pure oxygen yielding combustion products of water and carbon dioxide. Once dried, the flue gas would be carbon dioxide to high purity. The term "pipeline purity" is sometimes used (at least 95%), while amine scrubbing yields CO_2 of extremely high purity—around 99.9% pure [15]. As discussed in Chap. 6, high purity CO_2 can be an industrially useful material. An important consideration when planning CCS for an industrial process is to avoid, if at all possible, the dilution of any CO_2 gas stream. Of course a traditional unabated CCGT power station dilutes CO_2 at the top of the smokestack as the combustion products are vented to the atmosphere. At that moment, the CO_2 concentration drops from the flue gas concentration of around 11% to the ambient atmospheric concentration of 0.04%. Generally, we should be seeking to build a future based upon CO_2 concentration and storage, not CO_2 dilution and venting.

Linked to the idea that one should avoid ever down-blending CO_2 gas streams is the notion that whenever there is a choice: higher process concentrations of CO_2 are always to be preferred. Indeed, once one can achieve pipeline purities, a whole set of industrial opportunities open up. Indeed even today, in order to meet the needs of industrial processes, CO_2 is extracted from underground geological sources and piped to industrial facilities. The extraction of geologic CO_2 is frankly the exact opposite of what the world should be doing. As things stand, however, it makes business sense given prices and incentives. Related issues will be considered further in Chap. 6.

Higher CO_2 concentrations can be achieved earlier in the power generation process via a move to oxy-fuel combustion. In normal combustion, a hydrocarbon fuel, such as natural gas, is burned in air (usually at high pressure). As such, the exhaust gas comprises mostly nitrogen with some carbon dioxide (as discussed earlier). For oxy-fuel combustion, equivalent hydrocarbon fuels such as natural gas are burned in pure oxygen. Neglecting for now the role of minor impurities, this means that the exhaust gas comprises pure water and carbon dioxide. Once dried, the flue gas would be carbon dioxide to very high purity. Other benefits include in principle higher energy conversion efficiency, as for instance, there is no need to heat large amounts of nitrogen gas. Another benefit is the elimination of any risk of producing hazardous (NO_x) oxides of nitrogen. The process, at least as applied to large-scale power production, is not without its technical challenges. The main disadvantage has direct links to a key advantage—higher thermodynamic efficiency. Power generation thermodynamics are favoured by higher combustion temperatures, but the combustion of fuel in pure oxygen results in temperatures that are simply too high. By way of evidence, one can point to the use of pure oxygen in welding where the intention is to melt and bond structural materials, such as steels. The most straightforward and efficient way to reduce the combustion temperature is to dilute the fuel and oxygen and to increase mass flow through the combustion chamber. Ideally, one seeks a material that is immune to further oxidation and that is already at a high temperature for other reasons. The perfect candidate material to blend with the incoming fuel and oxygen and fuel is hot and dry carbon dioxide gas. The vast majority of oxy-fuel pilot plants and demonstrators adopt some variant of this approach. One variant under investigation is "staged combustion" in which combustion occurs in two stages. The first minor combustion generates relatively little energy but creates ambient combustion products; these then dilute the fuel and oxygen ready for the second and more important combustion stage. Such approaches are best suited to internal combustion engines and hence smaller implementations.

Proposals for enhanced oxy-fuel combustion would greatly increase demand for separated oxygen. It must be conceded that this might provide an eventual spur to Green Hydrogen production methods (both thermal catalytic splitting and high-temperature electrolysis methods [see Chap. 4]). It is one area where Green Hydrogen and Mature Hydrogen could drive the growth of a low-carbon hydrogen economy better together than in competition.

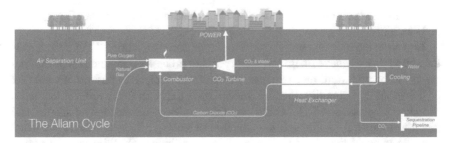

Fig. 5.4 Schematic of the Allam cycle developed by Rodney Allam and as being developed by NetPower in Houston Texas [16, 17]. *Source* NetPower

As regards the recirculation of oxy-fuel exhaust gases in power generation, one particularly attractive proposal is the Allam cycle named in honour of its developer, the British chemical engineer Rodney Allam. The Allam cycle is illustrated schematically in Fig. 5.4.

The Allam cycle is elegant because it requires only a single combustor and gas turbine. The concept completely avoids the use of water as an energy transfer fluid and hence avoids any inefficiencies associated with water phase transformations (such as recondensation). The Allam cycle does not consider CO_2 as an after-thought and CCS is not regarded as a modification of a prior power generation concept. Rather CO_2 and CCS are core fundamental attributes, if anything the need for efficient power generation is designed around the carbon management. There are plans underway led by North Carolina-based company NetPower LLC to commercialise the Allam cycle in La Porte, Texas close to the hydrogen infrastructures discussed in Chap. 6. The first step will be a 50 MW demonstration unit. Work on-site commenced in March 2016, and it is planned that the project will require investment of $140 million [16].

As it neared completion in 2017, the NetPower pilot plant in Texas started to secure some significant attention, such as from the influential Massachusetts Institute of Technology magazine *Technology Review*. In late August 2017, the magazine featured NetPower's ambitions and reminded its readers that, according to Julio Friedmann, Chief Energy Technologist at Lawrence Livermore National Laboratory, "If you keep the entire cycle above the supercritical phase, the efficiencies are amazing" [18]. The phase diagram of carbon dioxide is such that fortuitously that goal is achievable and far more easily than any equivalent or similar ambitions in water-based cycles.

Looking ahead a key challenge for NetPower, and others interested in oxy-fuel combustion, will be the prospects for improved air separation. In this book, we stress the production and supply of hydrogen from fossil fuel sources. If a hydrogen economy does take off, perhaps led by the international oil companies, one can expect it also to lift interest in hydrogen from other sources, such as water splitting (thermal and electrical—see Chaps. 1 and 4).

5.2   Carbon Capture Direct from the Air

Today in much of the word, we see a situation in which process chemical CO_2 is allowed to vent to the atmosphere without constraint. In this book, we have considered many ways in which such emissions might be reduced. There is, however, a more radical idea for atmospheric carbon reduction. It is the most ambitious form of CCS—the separation of ambient carbon dioxide from the atmosphere.

The term "air capture" refers to the capture of CO_2 from air. Air capture technology and its industrial application have been in existence, since as far back as 1930. There are three main pathways of air capture:

Direct air capture (DAC) uses filter/chemical process to capture CO_2

Enhanced natural sink (ENS) is executed by enhancing natural sinks (oceans, soil/vegetation) to capture to capture CO_2

Biomass coupled with CO_2 Capture and Storage (CCS) uses photosynthesis to remove CO_2 from the air.

However, ENS and biomass coupled pathways requires access to fertile land unlike the DAC pathway. Such considerations may limit the potential for these technologies [19].

Direct air capture is a form of negative emissions. The DAC process works by adsorption and desorption of CO_2; it involves fans blowing ambient air over capture material (termed the "filter"). The filter traps CO_2 and water and CO_2-free air is then released. The capture material is then reheated to $100^{O}C$. Water is released as a by-product, while CO_2 is released from the filter and collected as concentrated CO_2.

A DAC plant is sometimes referred to as artificial tree (because, like a tree, it absorbs CO_2 from the air). However, it is worth mentioning that there is a widely held perception that the DAC CO_2 process is extremely costly in both energy and financial terms, namely $600–$1000/tonne of CO_2 [20]. This estimate is based on extrapolating from knowledge of CCS obtained from fossil fuel combustion power plants. The CO_2 levels in power station flue gases are much higher than is found in ambient air. Hence, it is widely expected that the costs, and energy needs, for DAC will be greater than for more conventional power plant CCS and by implication even more unattractive when compared to the most viable capture option—process plant CCS. One of the companies that produce DAC plants has claimed that they could drive down the cost of production to $200/tonne of CO_2 in 3–5 years after developing their technology through a couple of iterations. They report a long-term goal of reducing the cost to just $100–$150/tonne of CO_2. Captured CO_2 from such processes can be stored underground to directly contribute to negative emissions and so help reduce global climate change.

DAC plant CO_2 might one day be used to produce ultra-low carbon intensity transportation fuels at an affordable price point using posited air-to-fuel technology (synthesising transportation fuels using only atmospheric CO_2 and hydrogen split from water, and powered by clean electricity)—this relates to the next section where we consider carbon utilisation. Such ideas have been advanced by proponents of Green Hydrogen. At this point, however, they appear far from economic viability, and there are far easier CCUS options available at scale.

Captured atmospheric CO_2 can be used to produce materials such as steel, fillers and coatings. It can also be used to produce chemicals such as plastics, chemicals and fertilisers [21, 22] in a spirit of carbon utilisation.

5.3   Carbon Utilisation

Carbon capture and utilisation (CCU) is a more recent concept than CCS. It envisages the use of captured CO_2 for conversion into useful industrial products including synthetic fuels, chemicals, plastics or cement. Captured carbon dioxide can also be used for food processing. CCU is now starting to receive serious attention in the global research landscape. Part of the reason may be due to the uncertainty of the long-term consequences, cost and risks of CCS (see Table 5.1) as well as the potential value of products from CCU. Many observers and commentators in Europe see little or no connection between CCU and CCS. Indeed, at the European Union level, CCS still tends to be a stand-alone topic. Proponents of the distinction would assert that in many ways the two technologies have little in common. Firstly, concerning costs, CCS is basically a waste management technology, where every step is costly and both public policy-makers and business strategists see CCS as a cost requiring support. CCU, on the other hand, has the potential to produce value-added products that have a market value and that can, in principle, generate a profit. Rather than focus on such distinctions, we tend to the view that there will be real merit in bringing these two realities closer together, although it will be a policy challenge. Secondly, those keen to stress difference point to the idea that the primary aim of CCS is to aid in the mitigation of climate change by storing large amounts of CO_2 underground. There is no inclination to add value to the captured carbon. In contrast, CCU's major driver is to substitute fossil carbon as a raw material for industry. That is it is to reduce the current need for carbon, not to seek to manage our carbon wastes. In such a mindset, CCU is akin to a recycling technology aimed at recycling CO_2. Whatever the rhetorical distinctions, CCU and CCS are closely related technologies with regard to carbon capture. We accept that CCU should not be limited as being just an enabler for CCS, as indeed it can do so much more than simply deposit carbon dioxide underground.

In some parts of Asia, CCU assumes greater importance due to the geological limitations for CCS. Currently, the use of CO_2 for Enhanced Oil Recovery (EOR) and Enhanced Coal Bed Methane Recovery (ECBMR) are the only options under consideration in many of the international reports. However, as these are not applicable in most of Asia. In the absence of fossil fuel related legacy systems, other utilisation options such as innovative reactions to convert CO_2 into useful products (fuels, value-added chemicals, building and construction materials, etc.) via organic, inorganic, or biological pathways are being explored, for example in Singapore.

The history of carbon dioxide utilisation actually long-predates the ideas of CCU as motivated by environmental concerns. Indeed, as noted previously, the USA has long operated carbon dioxide pipelines supplied by geologically sourced CO_2. Also as discussed further in Chap. 6, carbon monoxide (as sourced by the HYCO-POX process) is a valuable monomer, or building block, for the plastics industry. While plastic has come under close public scrutiny in 2018, it remains the case that recycling is far from sufficient to meet global needs for plastics production. Those aspiring to a wholly renewables-based future will need to confront the challenge of plastic supply given that the building blocks are currently provided as by-product of the fossil fuel industry. While the dominant focus may be on demand reduction and recycling, a third dimension deserves policy consideration—the role of plastics in CCU approaches and specifically the future role of fossil-fuel-based Mature Hydrogen in a low GHG emissions plastics production industry. This latter notion might be advance by greater use of Carbon Monoxide originating in HYCO-POX or SMR units. As discussed in Chap. 1, the future role of hydrogen will be dependent on the source and cost of hydrogen. Since the global imperative is to reduce CO_2 emissions, it is desirable to use a "Green" or a "CO_2 abated Mature" source of hydrogen. In European academic circles, we argue that disproportionate emphasis has been given to the former "Green" approach alone. We seek to stress that the latter Mature Hydrogen approach can have strong, and potentially powerful, synergies with CCU-based scenarios.

While SMR is indeed fossil-fuel-based, this book argues that its potential greenhouse gas credentials are easily misunderstood. For example, as introduced in Chap. 3, the first stage of steam methane reforming (SMR) yields carbon monoxide. Carbon monoxide is highly toxic and has geographically limited industrial uses. As a consequence, it is commonplace to include a "water-gas shift" reaction converting the carbon monoxide to more benign carbon dioxide, although it is a greenhouse gas. There are some important facts to summarise:

- Mature (fossil-fuel-based) hydrogen production methods are associated with carbon dioxide emissions for two reasons. The first is that the chemical engineering of the production process yields carbon monoxide (along with hydrogen). When processed through a water-gas shift reaction, the toxic CO is converted to more benign (but still environmentally harmful) CO_2. The second reason that Mature Hydrogen production is associated with CO_2 is the need for reagent heating to high temperatures. This is usually achieved through the combustion of the same compound that is the feedstock for the chemical conversion process, e.g. natural gas. To some extent at least, some of the heating role could be transferred to cleaner means (including even hydrogen, if incentives were sufficient to overcome the hydrogen production efficiency penalty that would be incurred).
- In some parts of the world, carbon monoxide is a valuable industrial reagent—it has been termed a *building block for higher molecules.*
- If CO is gas-shifted to CO_2, then one should note that the chemical process (as opposed to combustion for heating) yields a relatively high-purity CO_2 stream (approximately 45%) with only a small amount of hydrogen contamination. As such, this relatively concentrated CO_2 should not be confused with the tailpipe

emissions of a conventional fossil-fuelled IC engine or the smokestack of a coal-fired power plant. Those emitted gases are predominantly nitrogen. For these reasons, SMR water-gas-shifted CO_2 presents far fewer handling and management problems than smokestack emissions.

CCU technologies will play a major role in the future when it comes to adapting to a changing raw materials market—both in the energy sector and in the chemical sector. CCU can deliver solutions to major challenges looming for those seeking to advance renewables-based futures. Previously, we have considered such logics in connection with widespread electrification and Green Hydrogen. When seen from such a perspective, CCU can help to support the transition of the energy system towards fluctuating renewable energies, and CCU technologies can provide the means for large-scale energy storage with minimal land use requirements. The technology can also support the transition of the transport sector by providing technologies for clean fuel production from non-fossil sources with an extremely low carbon footprint. But with all that said, for those advocating a renewables-dominated future, the major contribution CCU is the possibility that it might help with the provision of an alternative raw material base for the chemical industry once the fossil fuel industry has been shut down. By developing CO_2-based production routes for base chemicals, the dependency on fossil carbon sources of the chemical industry and all subsequent production routes will decrease. Moreover, as an additional benefit, all these factors also help to mitigate greenhouse gas emissions significantly.

While generally in this book we do consider CCU as an attractive alternative to CCS, we also join others in welcoming it as an enabler of CCS. Others, however, seeing little of no merit in fossil-fuel-based futures (even low carbon futures), will often draw a significant distinction between CCS and CCU. While they might express enthusiasm for CCU, they frequently remain deeply averse to CCS. From the perspective of this book, we see much merit in both CCS and CCU, and we see many potential beneficial synergies in the broad space of CCUS. We are motivated by a deep concern for the global climate and focussed on the need to make the maximum beneficial change to emissions with the greatest haste and with the least social discomfort. We are concerned that to exclude meretricious ideas might reduce the prospect of success in addressing the climate challenge. We will return to such thoughts in Chap. 12.

5.3.1   Fracking and the Plastics Industry

At this point, it is appropriate to mention the role being played by US-led innovations in hydraulic fracturing technology (commonly known as "fracking") and the impacts seen for the plastics industry.

As mentioned elsewhere, the global natural gas industry has been profoundly reshaped over the last 15 years by largely unexpected developments in natural gas extraction. The USA's rush into fracking for natural gas extraction was so dramatic that it took the USA from a worsening position of gas dependence to gas surplus in just a few years. The consequential downward pressure on US natural gas prices was significant. As the fracking industry adjusted to new economic realities (fundamentally unexpectedly cheap gas), the technique found wider application in the extraction of tight oil, but it also led to the search for "wet gas". Schlumberger defines "wet gas" as "natural gas that contains less methane (typically less than 85% methane) and more ethane and other more complex hydrocarbons" [23]. While historically, a reduction in natural gas purity might have been regarded as a problem, in the case of fracked wet gas it is a positive attribute. Indeed, the industry is now actively seeking wet gas. The reason is the value of the non-methane hydrocarbons that can be obtained—they are extremely well suited to the needs of the chemical industry. In recent years, the US chemicals and plastics industry has seen a renaissance partly on the back of early national leadership in fracking technology [24]. While these developments are not directly related to a future hydrogen economy, they indicate to us how energy science and technology can take an unexpected turn and that such major shifts need not be driven by policy. We further posit that a CCUS revolution might throw up a similar set of industrial, and even ecological, benefits.

In essence, we suggest the innovativeness of the established oil, gas, industrial gases and chemical industries must not be underestimated when imagining the path to a low emissions future. While indeed they have surely been a major part of the problem over more than a century, this was in the service of a civilisation focussed on energy goals giving no thought to atmospheric consequences. Now climate change is rightly on the agenda—industry can, indeed must, be part of the solution. A major move into CCUS would be a key part of that.

5.4 The Way Ahead for CCS

Commercialising CCS is not simply a technical challenge. Indeed, it is arguably more a policy and regulatory challenge. It includes, for example, the need to establish incentives for early investment in CCS. It is noteworthy that in the twenty-first century, the UK has invested enormously more in new renewable energy system capacity than it has in CCS capability. This substantial funding difference reflects, in part, that CCS has not been afforded sufficient policy support, especially when viewed in terms of its ability to achieve deep CO_2 emissions reductions. If the approach based upon a cleaning up of Mature Hydrogen, production processes highlighted by this book is to be successful, then effective policies to accelerate the deployment of CCS (and looking ahead to CCUS) will need to be implemented within a decade.

CCS has gained momentum as a promising technology to facilitate GHG emissions reduction. Two commonly identified impediments to the widespread deployment of CCS include the cost of implementing CCS and a lack of regulation addressing unique CO_2 storage issues. While CCS development has already received offers of subsidy and benefitted from consideration of various financial incentives and political support, it remains necessary to develop a comprehensive and robust policy framework in order to give potential investors regulatory certainty and stable financial incentives. Comprehensive and broadly implemented legislation that puts a price on carbon would help encourage investment in carbon abatement technologies and help offset the current cost disadvantage of CCS. Well-designed regulations could mitigate the risks faced by investors by clearly identifying the ownership of CO_2, the scope of associated potential liabilities and remediation obligations, as well as the long-term liabilities relating to CCS.

CCS has started to gain momentum as a promising technology to facilitate GHG emissions reductions. While CCS has enjoyed various financial incentives and political support, it is equally necessary to develop a comprehensive legislative framework to give potential investors regulatory certainty and stable market-based incentives. Comprehensive and broadly implemented legislation that puts a price on carbon would encourage investment in carbon abatement technologies and help offset the current cost disadvantage of CCS. The second major type of impediment—an unclear regulatory environment—creates a risk of unpredictable and un-measurable liability that impedes investment. Well-designed regulations would mitigate this risk by clearly identifying the ownership of CO_2, the scope of associated potential liability and remediation obligations, and the long-term liabilities associated with CCS.

In summary, CCS is a key enabler of a potentially environmentally benign pathway for Mature Hydrogen production, distribution and use. The prospect of incorporating utilisation, as part of a CCUS vision, would further greatly boost this pathway to an economically viable and environmentally responsible hydrogen economy. Such issues will be considered further in the next chapter.

References

1. The EU Emissions Trading System (EU ETS). 2013. Available from: http://ec.europa.eu/clima/publications/docs/factsheet_ets_en.pdf.
2. Johanna, C., et al. 2015. Trends and projections in the EU ETS in 2015. In *European Environment Agency*, E.T. Report, Editor, Denmark.
3. The Costs of CO_2 Capture, Transport and Storage. European Technology Platform for Zero Emission Fossil Fuel Power Plants, January 4, 2019. Available from: http://www.zeroemissionsplatform.eu/library/publication/165-zep-cost-report-summary.html.
4. Gary, M.Z., ed. 2012. *Sustainable energy pricing: Nature, sustainable engineering, and the science of energy pricing*, 590. Wiley-Scrivener: New Jersey.
5. Knoope, M.M.J., A. Ramírez, and A.P.C. Faaij. 2015. Investing in CO_2 transport infrastructure under uncertainty: A comparison between ships and pipelines. *International Journal of Greenhouse Gas Control* 41: 174–193.
6. International Energy Agency, Five Keys to Unlock CCS Investment. 2017, January 4, 2019. Available from: https://www.iea.org/media/topics/ccs/5KeysUnlockCCS.PDF.
7. The Global Status of CCS. 2015. Available from: http://hub.globalccsinstitute.com/sites/default/files/publications/196843/global-status-ccs-2015-summary.pdf.
8. Steve, V. *Carbon tax cost killed off Scottish Power's Longannet CCS plan*. Sunday Herald 2013. Available from: http://www.heraldscotland.com/business/13088856.Carbon_tax_cost_killed_off_ScottishPower_s_Longannet_CCS_plan/.
9. Haszeldine, S. 2018. Perspective: CCS—Aiming for Gold at Rainbow's End. In *Petroleum Review*.

10. Smit, B., A.-H.A. Park, and G. Gadikota. 2014. The grand challenges in carbon capture, utilization, and storage. *Frontiers in Energy Research* 2 (55).
11. Coninck, H.D. and S.M. Benson, carbon dioxide capture and storage: Issues and prospects. *Annual Review of Environment and Resources* 39 (1): 243–270.
12. CCGT Technology. January 3, 2019. Available from: http://www.marchwoodpower.com/ccgt/.
13. Zevenhoven, R. and P. Kilpinen, FLUE GASES and FUEL GASES. In *Control of pollutants in flue gases and fuel gases*. Finland: Espoo/Turku.
14. Rochelle, G.T. 2009. Amine scrubbing for CO_2 capture. *Science* 325 (5948): 1652–1654.
15. Amine Scrubbing Process. 2018. Available from: https://hub.globalccsinstitute.com/publications/final-report-project-pioneer/amine-scrubbing-process.
16. NET Power Breaks Ground on Demonstration Plant for World's First Emissions-Free, Low-Cost Fossil Fuel Power Technology. 2016. Available from: https://www.prnewswire.com/news-releases/net-power-breaks-ground-on-demonstration-plant-for-worlds-first-emissions-free-low-cost-fossil-fuel-power-technology-300233131.html.
17. Roberts, D. 2018. That natural gas power plant with no carbon emissions or air pollution? It works. Available from: https://www.vox.com/energy-and-environment/2018/6/1/17416444/net-power-natural-gas-carbon-air-pollution-allam-cycle.
18. Temple, J. 2017. Potential carbon capture game changer nears completion, May 2, 2018. Available from: https://www.technologyreview.com/s/608755/potential-carbon-capture-game-changer-nears-completion/.
19. Ranjan, M., and H.J. Herzog. 2011. Feasibility of air capture. *Energy Procedia* 4: 2869–2876.
20. Evans, S. The Swiss company hoping to capture 1% of global CO_2 emissions by 2025. Clear on climate 2017; Available from: https://www.carbonbrief.org/swiss-company-hoping-capture-1-global-co2-emissions-2025.
21. Direct Air Capture. 2018. Available from: http://carbonengineering.com/about-dac/.
22. Magill, B. *World's first commercial CO_2 capture plant goes live.*. Available from: http://www.climatecentral.org/news/first-commercial-co2-capture-plant-live-21494.
23. Schlumberger, Oilfield Glossary 2019 January 4, 2019. Available from: https://www.glossary.oilfield.slb.com/en/Terms/w/wet_gas.aspx.
24. With thanks to Charles Forsberg for inspiration in Private Communication 2018 (All responsiblity lies with the authors).

Chapter 6
Hydrogen Infrastructures

6.1 Hydrogen Storage and Distribution

As noted in Chap. 1, while globally the hydrogen market is competitive with a range of highly capable industrial gases companies present in the market. In some territories, there is a risk of market power or even market dominance. At the risk of over-simplification, microeconomic theory suggests that if a market has fewer than five sellers, even without any illegal collusion, competition will be imperfect. The extreme limit of such market power would be one sole vendor. This would represent a monopoly—a situation that allows the vendor sufficient market power to charge prices substantially higher than would be seen in competitive markets. The restrictions and restraints on supply can also be physical in origin, such as a lack of proper distribution and transportation infrastructure and limited availability of hydrogen storage technologies. These capability gaps can be difficult and expensive to fill and, as such, represent a barrier to entry for newcomers. A widely occurring obstacle to the wider adoption of hydrogen for energy applications is its consistently high cost per unit of useful energy when compared with the precursor fossil fuels used in its manufacture. An appropriate emissions charge for CO_2 would militate against that disparity and encourage, for example, the use of carbon capture and storage with Mature Hydrogen, or the production of renewable-based Green Hydrogen. If we imagine that the future costs of Green Hydrogen might indeed be low, policy-makers should remember to look to the wider social impacts that may be arising from the socialisation of costs, such as those underpinning renewable support in electricity generation.

Looking ahead to future challenges, the coupled deployment of hydrogen storage and distribution is particularly difficult. It can be regarded as an example of the "chicken and egg" problem, as each (storage and distribution) requires the other to pre-exist it. Breaking through in this domain will require significant technology innovation, policy intervention and business initiative.

Hydrogen has an attractive attribute in that it is well suited for relatively long-duration energy storage. For instance, it has the potential not only to smooth diurnal

© Springer Nature Switzerland AG 2020
W. J. Nuttall and A. T. Bakenne, *Fossil Fuel Hydrogen*,
https://doi.org/10.1007/978-3-030-30908-4_6

fluctuations in electricity generation and supply, but it has the potential to address the much more challenging goal of seasonal storage. The relative attributes of hydrogen storage are summarised in Fig. 6.1.

The generation and delivery of hydrogen is a particularly rich whole system problem. It is a network problem in those elements of the system have little, or no, value in isolation. One key system choice is whether to produce hydrogen close to the point of use (or retail sale) or to rely on more centralised approaches. If not locally produced, then it must be distributed as a cryogenic liquid or high-pressure gas by truck or as a gas through a pipeline. As seen in Fig. 6.2, the supply chain economics is set by the distances involved and the scale of demand. At small-scale and over

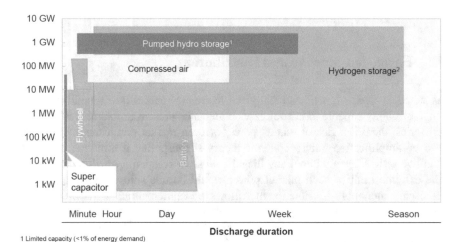

1 Limited capacity (<1% of energy demand)
2 As hydrogen or SNG

Fig. 6.1 Relative energy storage capabilities of various energy storage options. *Source* Hydrogen Council (hydrogencouncil.com) [1]

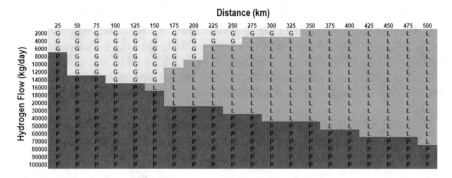

Fig. 6.2 Mode map showing the lowest cost options for hydrogen delivery. Results are presented in terms of hydrogen flow and kilometres of transport distances, where "G" stands for gas trucks, "L" for liquid hydrogen trucks and "P" for pipeline gas supply [2, 3]

shorter distances, the lowest cost method for production, storage and distribution is as a compressed gas.

The merchant producers serve and deliver gases for industry, health and government through various means, including pipelines, tube trailers, bulk and cylinders, small on-site plants and large on-site plants. Gas pipelines are at present the most feasible option for transmitting large quantities of hydrogen over long distances (see Fig. 6.2), and pipeline transmission is expected to be the means by which hydrogen is delivered from future large-scale, centralised production plants and distributed to fuelling stations [4]. However, barriers to the pipeline approach include high capital costs and materials engineering issues, such as the purported hydrogen embrittlement of pipelines. The existing hydrogen pipeline technology and network are not enough to meet demands and consequently cannot achieve the cost and performance goals required for successful implementation of a commercial distribution network. Another major hurdle is related to the fact that many economic actors will need to take co-ordinated action if hydrogen is to achieve a significant share of the market for transport fuel. Customers will not purchase fuel cell electric vehicles (FCEVs) unless there are sufficient fuelling stations; car manufacturers will not invest in producing vehicles that people will not buy, and fuel vendors will not install hydrogen stations for vehicles that do not exist.

Air Liquide, Air Products and Linde AG are the main hydrogen production companies in Europe. These industrial gases companies support a hydrogen pipeline network, as shown in Fig. 6.3. In this chapter, we shall focus on the European operations of Air Liquide (AL) and the operations in the USA on the Gulf Coast operated by Air Products (AP).

6.2 European Experience

Air Liquide (AL) is the largest hydrogen producer in Europe. 66% of the hydrogen sold by the AL group is used for vehicle fuel desulphurisation (hydrotreatment). AL operates 12 pipeline networks worldwide that together cover more than 1850 km. The longest pipeline is located in an area spanning Northern France, Belgium and part of Netherlands, covering 1100 km. The other major pipeline is located along the Gulf of Mexico and the USA and is operated by Air Products. AL has ~50 hydrogen production units across the globe, producing 0.89 million tons of hydrogen [6]. AL hydrogen solutions are used to prevent the release ~780,000 metric tons of sulphur oxides/year [6]. This is a substantial quantity representing twice the total sulphur emissions of a country as large as France (Fig. 6.4).

Hydrogen is usually encountered as a very light gas, and one can think of its use in early balloons and airships. At atmospheric temperature and pressure, 1 kg of hydrogen gas occupies more than 11,000 l. For practical storage, one needs to compress the hydrogen molecules closer together. At a pressure 700 times that of the atmosphere (i.e. 700 bar), 1 kg of hydrogen occupies 23 l. Further densification can

Fig. 6.3 Northern Europe hydrogen pipelines. *Source* Wiley with permission, image from [5] ©
Air Liquide

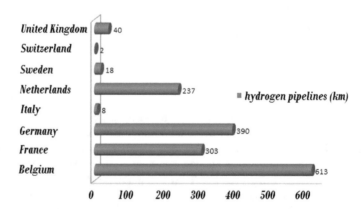

Fig. 6.4 Major hydrogen pipelines network across Europe. Adapted from [7, 8]

be achieved by liquefaction, but this is not possible at room temperature. If liquefied
at −253 °C (20 K), then 1 kg of hydrogen can be transported in a 14-l container [6].
Some of these ideas were introduced in Chap. 2, and the issues surrounding liquid
hydrogen will be discussed further in Chap. 9. Industrial gases companies, such as Air
Liquide, have developed facilities for storing high-pressure (350–700 bar) hydrogen
gas and also cryogenic liquid hydrogen (LH$_2$). For those customers requiring only

modest quantities of hydrogen, delivery is made in gas bottles or in tube trailer trucks where hydrogen can be stored in compressed gas form. Since 2002, Air Liquide has designed and built more than 100 hydrogen filling stations worldwide. Since 2012, Air Liquide has been providing such services to serve the general public. These stations allow vehicles to refuel fully in less than 3 min with the same safety required by a conventional petrol station.

The design and operation of hydrogen pipeline infrastructure come with challenges. Hydrogen is a colourless and odourless gas and any small leaks in the pipeline are hard to detect. Any hydrogen pipeline brings the possibility of leak-related fires. As such hydrogen pipelines require constant monitoring and ongoing maintenance. This level of attention exceeds even that given to more conventional pipelines (such as for natural gas). The extra attention comes at a price and this can erode the economic competitiveness of some hydrogen-related business propositions; at least as things stand at present.

The industrial gases companies generally and, Air Liquide in particular, have aimed to reduce CO_2 emissions significantly in the years after 2010. This Air Liquide plans to do by producing 50% of their hydrogen using renewable energy and by exploring the use of steam methane reforming with carbon capture and storage (CCS). However, as discussed in Chap. 5, the moves towards CCS have been held back by the generally very low greenhouse gas emissions prices seen in the European Union under the EU Emissions Trading Scheme. The EU-ETS price has consistently been far below what is needed for CCS to be economically rational given the high costs of the technology. Sadly, it is simply far cheaper to emit. Only in 2018 has the EU-ETS price started to pick up, but even in the range 20–30 Euro per Tonne of CO_2, it remains insufficient. It is not technology that is holding back environmental progress, it is energy policy and carbon market design.

6.3 The US Gulf Coast

As one looks around the world for examples of a "Big Hydrogen" industry, candidate locations include Benelux regions connected to hydrogen infrastructures around the Port of Rotterdam (discussed in the preceding section) or perhaps as can be seen in Singapore. In future, China offers much similar potential. The clearest current example, however, is the Gulf Coast region of the USA as served by the Air Products Gulf Coast Connection Pipeline stretching more than 180 miles from the Houston Ship Canal in Texas to New Orleans, Louisiana [9], see Fig. 6.5.

The key attributes of the Air Products' hydrogen pipeline infrastructure are its scale and completeness. This gives rise to important collective benefits allowing synergies between the numerous elements collected to the pipeline. In addition, the pipeline itself serves as an important storage facility for hydrogen—so-called "line packing". Assets continue to be added to the pipeline with four large hydrogen production plants having joined the pipeline over recent years (Baton Rouge, Garyville, Luling and NORCO).

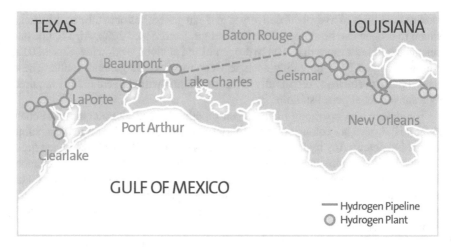

Fig. 6.5 Air Products' hydrogen pipeline runs for more than 180 miles along the US Gulf Coast. *Source* From [11]

It is vital to stress the importance of a diversified set of producers and users in the vicinity of the pipeline. Users include those concerned with oil refining operations and those involved in chemical processing (polymerisation for plastics). Production includes both SMR and POX-based HyCO plants. As such, the hydrogen pipeline is surrounded by other important synergistic infrastructures such as the Baytown to Freeport carbon monoxide pipeline. In addition, the region has carbon dioxide pipelines, some supplied with geological CO_2 and others, such as the Texas Green Pipeline for Enhanced Oil Recovery associated with CO_2 flows back underground (this pipeline has innovative CO_2 handling technology).

The role of the La Porte HyCO III plant in this extended infrastructure is particularly interesting and important. Fundamentally such a facility, in such a location and with the supporting infrastructure, has four products. They are:

1. Hydrogen
2. Steam
3. Electricity
4. Carbon monoxide.

It is important to stress that the infrastructure (including the hydrogen pipeline) brings the four commercial products together. In such a context, the role of CO manufacture is extremely important and valuable. "Steam hosts" are also important— i.e. *who will buy my steam?*

While hydrogen is indeed an important product, its importance is not so overwhelming. For some "HyCO" processes around the world, it is very much a by-product and indeed given the industrial interest in the other three products, and the hydrogen is not so far from being seen as a waste.

We have stressed the importance of POX-HyCO assets when connected to an hydrogen pipeline. While that is true, the Air Products experience on the US Gulf

Coast shows that a pipeline is not essential in order for some Big Hydrogen (see Chap. 8) benefits to be obtained. For example, the Air Products Gesmer facility is an isolated CO production asset unconnected to any CO pipeline, but the CO is still a merchant product transported by tube trailer. For merchant CO and H_2 production: tube trailers are a relevant near-term capability. This experience could be highly relevant to the emergence of Big Hydrogen systems elsewhere in the world. The main requirement is the demand for industrial CO feedstock gas, rather than a full pipeline infrastructure as seen on the US Gulf Coast.

Interestingly today, the La Porte III POX facility runs on natural gas, for fuel and feedstock, but there is a history of it running as a POX gasifier fuelled by Exxon Mobil's "Bayton" tar. Tar gasification may seem a hugely CO_2 intensive process, but for reasons that we hope which are becoming clear even that assumption is not necessarily so obviously true.

Mature Hydrogen today has much to do with the needs of hydrogen production to serve customers in the petroleum refining business. As such it is intimately connected with the petroleum-based, and CO_2 emitting, traditional paradigm for transport and mobility based on the internal combustion engine. In fact, however, the Mature Hydrogen infrastructure above is technologically neutral. It can serve any customer that might desire the hydrogen produced.

On the supply side, we have seen the potential to innovate incrementally towards a low CO_2 future simply in response to pressure from relatively transparent market-based incentives. If such incentives were wide-ranging (e.g. a pervasive and escalating price CO_2 price), then we would expect gradual demand shift from petroleum uses to direct uses of hydrogen as a fuel.

To understand the early stages of such a shift, and with reference to the US Gulf Coast infrastructures, one should note that Air Products has both its Operations Service Centre and its Customer Services Centre adjacent to one another in Houston near the hydrogen pipeline. These capabilities will very soon, perhaps already, permit full-service delivery from technical operations to hydrogen trading. Such a capability would allow, very easily, for the addition of a downtown Houston retail hydrogen point of sale to a vehicle fleet operator or even individual car owners. Such a step is technologically almost insignificant when compared to the investments already made in the Gulf Coast hydrogen infrastructure. Of course, any such connection to the hydrogen pipeline would be a small new user on the network, but both the pipeline and the business/trading model can cope easily with small additional users. In this way, we can imagine the roll-out of an environmentally benign new basis for transport and mobility, not from European centres of enthusiasm for Green Hydrogen, but rather from Houston Texas—the world's hydrocarbon energy capital.

Indeed, Houston's first hydrogen filling station opened on 5 December 2012 developed by Air Products with assistance from the Environmental Defense Fund. The station is fed with hydrogen from the Air Products' hydrogen pipeline [10].

6.4 Partial Oxidation—A Provider of Hydrogen and Molecular Building Blocks

Perhaps, the greatest opportunities relating to the expansion of existing hydrogen infrastructures will relate to the use of POX-HyCO production systems (see Chap. 3). The reason here is that once started, the POX process is exothermic. It does not need to be heated—it is self-heating greatly avoiding the need for fuel (e.g. natural gas) combustion. The other reason is that it is optimised to be a CO production system. Typically, there is no water-gas-shift reaction to CO_2. The CO is not a waste to be vented as CO_2, it is a product to be sold. If it is later converted to CO_2 (which seems rather unlikely), it would be an emission associated with someone else's industrial process. Far more likely, it will be used in the chemical process industry to make solid plastics, as already occurs in the US Gulf of Mexico cluster. These plastics might be recycled in the short term before later being long-term sequestered (via unprocessed disposal) or be combusted in a waste-to-energy plant. Such decisions will involve a wider set of environmental considerations than just greenhouse gas emissions, but it should surely be the case that any resulting emissions are not the responsibility of those merely producing a CO molecular building block. We therefore see that while an initial assessment might suggest that POX processes are environmentally problematic because of CO_2 emissions. In some industrial contexts, such an opinion could be largely erroneous as a result of an overly narrow consideration of the issues.

We must concede that the potential for HyCO-POX carbon monoxide and hydrogen joint production is probably limited to just a few centres globally. This is because there are relatively few places with substantial chemical process industry demand for carbon monoxide (CO). As noted earlier, CO can be thought of as a molecular building block for polymers and plastics. The Gulf of Mexico region is world-leading in such initiatives with the established HyCO-POX plant and an extended Air Products (and partners) infrastructure capable of linking four linked products, namely hydrogen, steam, electricity and carbon monoxide, all as commercial propositions. In those places where there is such demand, such as the US Gulf Coast, one can easily see an innovative path ahead for Mature Hydrogen production innovating towards relatively low environmental impacts incentivised merely by a rising CO_2 emission price and largely free of dependence on policy-based technology-push or for a socialisation of roll-out costs across the energy system.

6.5 US CO_2 Pipelines

As noted in Chap. 5, it is noteworthy that the US Gulf Coast region has a developed CO_2 pipeline infrastructure transporting geologically sourced CO_2. The existence of such CO_2 extraction and transport operations is almost the opposite of what might be expected in a carbon-conscious world. The existence of this established CO$_2$ infrastructure reveals local demand for CO$_2$. In future that demand might be met by supply from Mature Hydrogen production facilities allowing the geological CO$_2$ to remain underground (see Chap. 5). In summary, advanced SMR hydrogen production has the potential to reduce overall GHG emissions—in a phased manner (first substitution for the use of geologically sourced CO$_2$ then possibly a later move to CCS if CO$_2$ production exceeds industrial demand).

References

1. Hydrogen scaling up-a sustainable pathway for the global energy transition. 2017. Available from: http://hydrogencouncil.com/wp-content/uploads/2017/11/Hydrogen-scaling-up-Hydrogen-Council.pdf.
2. Yang, C., and J. Ogden. 2007. Determining the lowest-cost hydrogen delivery mode. *International Journal of Hydrogen Energy* 32 (2): 268–286.
3. Shayegan, S., et al. 2006. Analysis of the cost of hydrogen infrastructure for buses in London. *Journal of Power Sources* 157 (2): 862–874.
4. Knoope, M.M.J., A. Ramírez, and A.P.C. Faaij. 2015. Investing in CO$_2$ transport infrastructure under uncertainty: A comparison between ships and pipelines. *International Journal of Greenhouse Gas Control* 41: 174–193.
5. Ausfelder, F., et al. 2017. Energy storage as part of a secure energy supply. ChemBioEng Reviews 4.
6. Group hydrogen business figures. 2015. Available from: http://www.uk.airliquide.com/en/solution-for-industry/innovative-hydrogen-solutions-for-clean-energy/key-figures.html.
7. Hydrogen Analysis Resource Center. Energy efficiency and renewable energy 2015. Available from: http://hydrogen.pnl.gov/.
8. Maisonnier, G., Steinberger-Wilckens, R., Trumper, S.C. 2007. European hydrogen infrastructure atlas and industrial excess hydrogen analysis. Part II: Industrial surplus hydrogen and markets and production. In Roads2HyCom.
9. Hydrogen an evolving paradigm shift in Gasworld. 2013, 38–40.
10. Fernández, J.R., and J.C. Abanades. 2017. Optimized design and operation strategy of a CaCu chemical looping process for hydrogen production. *Chemical Engineering Science* 166: 144–160.
11. Air Products' U.S. Gulf Coast hydrogen network. 2012. Available from: http://www.airproducts.com/microsite/h2-pipeline/pdf/air-products-US-gulf-coast-hydrogen-network-dataSheet.pdf.

Chapter 7
The Proposed Natural Gas to Hydrogen Transition in the UK

In the authors' opinion, UK energy policy is shaped by a set of three policy concerns sometimes known as the "energy trilemma". The three concerns are: energy economics (primarily affordability), environmental responsibility (with a dominant concern for climate change mitigation) and security of supply (ranging from primary fuel security, through energy conversion capacity and including reliability in transmission and distribution). The term "trilemma" is a deliberate pun building upon the notion of a dilemma—i.e. a difficult choice between two options with little, or no, scope for compromise. In invoking the notion of a "trilemma", the implied meaning is clear.

For policy-makers, it would be straightforward to construct an energy policy to deliver on just one of the three concerns, it would not be so problematic to deliver on two (at the expense of failure in the third aspect) but to deliver all three goals simultaneously (i.e. energy that is low-emission, secure and cheap) is seemingly impossible. As a consequence, British energy policy must adopt a pragmatic portfolio approach in order to achieve an acceptable outcome that is likely to exhibit some levels of sub-optimality when seen in narrower terms (Fig. 7.1).

The authors have the impression that the UK government is sincerely committed to a long-term reduction in the emission harmful greenhouse gases, such as carbon dioxide arising from fossil fuel combustion. Indeed, the UK was the first country in the world to enact legislation mandating substantial reductions in such emissions [1]. Figure 7.2 illustrates the nature of the 80% emissions reduction task that the UK has determined must be met. While the details of how the goal (80% of net annual 1990 greenhouse gas emissions to be eliminated by 2050) is not prescribed by policy, the destination is indeed fixed in law under The Climate Change Act (2008) [1]. Figure 7.3 shows one way in which the goal may be achieved as suggested by the UK Climate Change Committee, a high-level advisory body established by the 2008 Act.

© Springer Nature Switzerland AG 2020

W. J. Nuttall and A. T. Bakenne, *Fossil Fuel Hydrogen*,
https://doi.org/10.1007/978-3-030-30908-4_7

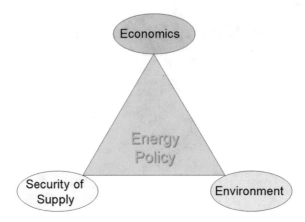

Fig. 7.1 The Energy Trilemma

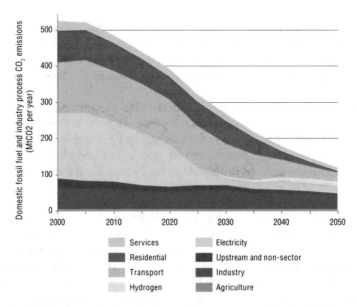

Fig. 7.2 One 80% carbon emission reduction pathway, as proposed by the UK Climate Change Committee. Services is the component at the top of the figure, and subsequent components follow the sequence in the key. Image with permission of OUP, *Source* [2]

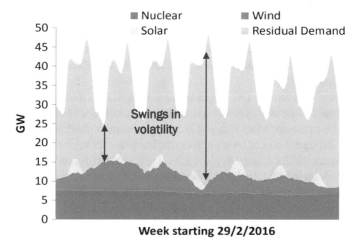

Fig. 7.3 Challenge of UK electricity system volatility. *Source* Alaric Marsden FTI Consulting at EPRG Spring Seminar May 2018 with permission, underlying data is obtained from National Grid's Electricity System Operator [3]

7.1 An All-Electric Future?

The pathway illustrated in Fig. 7.2 places emphasis on the decarbonisation of electricity generation. This is then followed by an expansion of electricity utilisation, especially in transport and mobility. As authors of this book, however, we propose an alternative way ahead. Yes, there must be a move to decarbonise electricity, but we suggest that a major increase in electricity usage (for example, into mobility and heating) could prove excessively costly. If one looks at the decrease in electricity emissions shown in Fig. 7.2, one gets no sense of the increase in electricity associated with a major electrification programme. This is because expansion will be largely on the basis of low, indeed near-zero, emission technologies.

We further posit that a major expansion of the electricity sector could align poorly with the interests and expertise of many important energy companies and it risks stranding significant and valuable current assets including those associated with natural gas processing, transmission and distribution. A low-carbon hydrogen-based vision for mobility and heating could be significantly easier for a country, such as the UK, to achieve if based to a significant extent on natural gas. It would require an evolution of current infrastructures associated with today's production and distribution of natural gas as well as an adjustment to what is today the logistics and retail sale of petroleum products. These activities would evolve as a consequence of a low carbon transition, but they would not be largely ended as they might be in a fully electrification-based scenario.

Part of the attraction of the electrified future, bringing mobility and electricity grids together is that problems, that today exist separately, might be eliminated collectively for mutual benefit. The risk, however, is that by bringing two tricky issues together (mobility and electricity) things might actually get worse, rather than better. The simplest manifestation of a worsening situation would be if electric mobility (and electric heat) were to increase stress on the electricity system rather than diminish it. Issues of large-scale electrification as an alternative way ahead will be considered further in Chap. 12.

The possibility that the UK might make a significant move into hydrogen (as opposed to extra electricity system expansion) brings a further political benefit which can briefly be summarised as follows: British government ministers like to make a substantial difference and to be seen to do so. Imagine, for sake of argument, that a minister responsible for energy deploys £500 million to help decarbonise electricity, then one can imagine that it might be perhaps the 25th step in a long policy journey, and furthermore, it might help industry make a 1% difference to overall national emissions. Fundamentally, the decarbonisation of electricity is already underway, and in policy terms, it is largely a done deal, with only marginal gains left to play for. In contrast, the decarbonisation of natural gas would be a new and dramatic proposition with the potential to make a truly material difference to overall national emissions. The allocation of £500 million to help push forward progress in that area would make a substantial overall difference and furthermore have a novelty and radicalism to ensure visibility for the policy within the government and externally. It would potentially have great impact on the national media and hence with voters. With such ideas in mind, a natural-gas-based way ahead would appear to be both effective and politically attractive, combining low carbon credentials with the potential for relatively attractive economics. The plan could be based on industry-led incremental upgrading of a mature set of infrastructures. Few assets would be left entirely stranded and few large well-established companies would be threatened. One must not neglect the reality that many of today's large oil and natural gas companies feature prominently in institutional investment portfolios and hence in the individual pension plans of voters. While some adjustment and risk management has already started to occur anticipating a low carbon future, it remains the case that these fossil fuel companies are extremely important to national prosperity. We shall also return to these ideas in Chap. 12.

The issues shaping the future of the electricity system in the UK are complicated and difficult even before one considers extending the reach of the system into areas of mobility and domestic heating. The system is already seeing growing swings in overall volatility driven largely by the growth in intermittent renewables on the supply side, see Fig. 7.3.

Figure 7.3 presents the swings in volatility seen over one week in the UK. One can clearly see the residual unmet demand after the contribution of the key low-carbon sources of electricity (nuclear and renewables) in February 2016. Hence, the figure shows the substantial role currently played by fossil fuel generation in meeting the residual demand. The figure also shows the significant daily, or diurnal, variation with a double-peaked demand. The highest demand of all being seen in the evening (typically at around 6 pm). Advocates of renewables point, with some justification, to the possibility of electricity storage helping to balance these diurnal variations in both demand and renewable energy supplies. Two issues shape the likelihood of this being

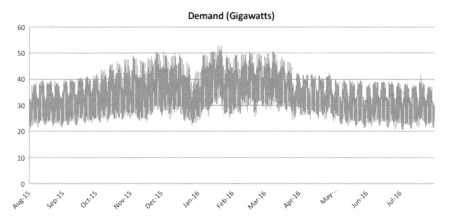

Fig. 7.4 Annual UK electricity demand authored by Chris Goodall [4]

possible: the scale of the challenge (i.e. storage capacity) and the duration of storage. As Fig. 7.3 reveals, the scale of the storage challenge to meet winter demand could be enormous. If very large-scale storage is developed, then the challenge would not simply be diurnal, but would actually be seasonal. It is the long-timescale duration of seasonal storage that is especially difficult to provide at reasonable cost today.

As Fig. 7.4 illustrates, the short-term diurnal oscillations in electricity are part of a wider seasonal variation. UK winter average electricity demand can be roughly one-third higher than an equivalent summer average. With today's still largely fossil-fuel-based electricity system, these differences are met by varying the use of fossil-fuelled generation technologies with relatively large marginal fuel costs. In essence, fossil fuel power plants are used intensively in the winter months and are often left largely idle in the summer months. However, in a largely decarbonised system, based on renewables with nuclear power, the marginal costs of generation would be very low (the fuel is very cheap or non-existent). Consequently, once the low-carbon generation sources exist, then they will always run if they can (as it costs essentially nothing for them to do so). Given the expected intermittency of most renewable power generation and given the winter peak in demand, it is not realistic to imagine that these levels of demand will be met simply from intermittent renewable energy capacity additions, and hence, storage would appear to be essential. An electrified future would require long-term seasonal electricity storage capacity as well as diurnal capacity if the total amount of installed generation capacity is to be kept within sensible bounds, but then it becomes very clear that the challenge inherent in such ideas for large-scale decarbonisation via renewable-based electrification is frankly daunting.

Contemplating such a future, it is clear that technical breakthroughs are still required. While there has been substantial progress in electricity storage technologies (especially for the diurnal challenge), the seasonal challenge remains uneconomic and seems currently to be far from achievable, although hydrogen might have a role to play (see Fig. 6.1).

As regards short-term storage systems, the proposal from Tesla for the *Tesla Wall* system is a good example of innovation supporting consumer-led initiatives. These home electricity storage units draw upon Tesla's extensive experience in battery electric vehicle (BEV) development. It is widely argued that BEVs will have the potential to help balance the future electricity system. Future real-time price incentives coupled with smart metering might favour charging at times of day most advantageous for system balance as opposed to the natural tendency that a motorist might to charge the car immediately on return from work (actually in the UK that is generally the worst time of day to do such a thing). In addition, it is widely hoped that one day car batteries might be connected to the local distribution grid such that they might be discharged providing power to the grid at times of system stress. Such ideas are attractive and radical, but also complicated. The Tesla Wall home installed battery for individual householder use is an idea that would appear to be more immediately applicable than notions of future vehicle-based storage for grid balancing.

Proponents of an electrified future point to the potential benefits to be obtained from "smart systems" involving greater use of information technology and the beneficial role that might be played by demand management in the electricity system. Thus far, however, the expensive large-scale programmes of the roll-out of "smart meters" are proving controversial with allegations that they have negligible beneficial impacts. The concerns have prompted the UK government to commission an updated cost-benefit analysis of the smart meter roll-out programme with the results expected to be made public in the summer of 2019 [5]. The domestic smart meters programme provides consumers with real-time energy use information, but so far real-time domestic consumer pricing in the UK remains rare, although Octopus Energy has launched a tariff known as Agile Octopus which leads the way in such innovation. Generally, however, the vast majority of end-user consumers (even those with smart meters) do not yet see the short-term price fluctuations seen in the wholesale market and hence have no ability or incentive to engage with such issues. While much has been made about (admittedly disputed) impacts of smart metering on domestic energy demand and the consequent reductions in consumers bills, it is perhaps easier to expect that the information benefits for suppliers might be proving useful in terms of system balance planning (at all levels). It may be these latter benefits that truly justify the smart meter roll-out programme, but such benefits are hard to assess or monetize, especially for industry outsiders.

It is far from clear that it has been appropriate to socialise the costs of the domestic smart meter roll-out across electricity consumers if the principle impacts are private (and invisible) benefits to the electricity companies. The fact that smart meter enthusiasm had to wait for a national policy push, rather than having occurred spontaneously as an industry initiative as soon as the digital technology became available, rather implies that the narrow benefits to the electricity companies are insufficient to motivate domestic deployments.

As authors of this book, we see much potential for smart-enabled demand management, electricity storage and new renewable generation to combine to deliver a

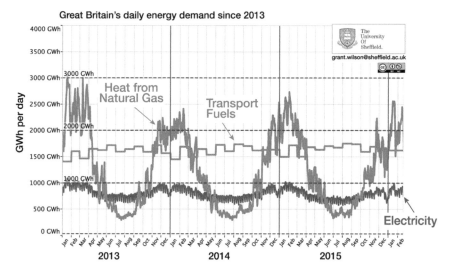

Fig. 7.5 UK energy system as a whole—showing seasonal and shorter-term variability [6]. *Source* Grant Wilson, University of Sheffield, UK under CC-BY-NC Licence

reliable low-carbon electricity system. We further posit that an expansion of nuclear power might make this challenge easier to meet, but we regard that as a secondary matter. Our concerns start to grow however when the proposition becomes the notion that these approaches might be extended as part of electrification of mobility and domestic heat. As Fig. 7.5 makes clear, the scale of British transport and heat challenge are such that electricity cannot easily be a major part of the solution to that problem. Rather than seek to displace natural gas, we suggest that the goal should be to seek to clean up mobility and heat by other (non-electrical) means and separately allow the electricity system to decarbonise within its existing, and already challenging, goals.

Figure 7.5 demonstrates the scale of the seasonal whole energy system challenge for the UK. Transport energy demand lies far above current electricity supply levels. The challenge presented by UK domestic heating is even more daunting. It seems quite impossible that this demand could be met by electricity-based measures irrespective of the need for an electricity system transition to a low carbon future.

7.2 A Future Role for Hydrogen in UK Domestic Heat

In this book, we suggest that another way must be found if heat and mobility are to be successfully decarbonized by mid-century. We concede that a British future based upon a natural gas to hydrogen transition would be ambitious. It would, however, make a genuine and material difference to total UK greenhouse gas emissions. Perhaps most importantly, it is a proposition well-suited to UK technological capabilities (both in terms of assets and experience).

We are not alone in seeing a compelling way ahead based upon extensive use of natural-gas-sourced clean hydrogen in the UK. Thus far, much of this book has focussed on the important challenge of decarbonising global mobility, but it gives us pleasure to observe that UK thinking is currently somewhat bolder than even that—with major ambitions linking hydrogen with both mobility and urban heating. We note, for example, that the *UK Clean Growth Strategy* has an explicit Hydrogen Pathway [7].

That document states (p. 56):

> Under this pathway, we use hydrogen to heat our homes and buildings, as well as to fuel many of the vehicles we drive in 2050 and power the UK's industry. We adapt existing gas infrastructure to deliver hydrogen for heating and a national network of hydrogen fuelling stations supports the use of hydrogen vehicles. A large new industry supports hydrogen production using natural gas and capturing the emissions with CCUS.

Further details concerning the Hydrogen Pathway are given on page 151 of the Clean Growth Strategy:

> All cars and vans are fuelled by hydrogen and the majority of buildings use a hydrogen grid. Electricity and district heat still play a role in both residential and commercial/public buildings. Overall hydrogen production is around 700 TWh in 2050, with Steam Methane Reforming and CCUS being the primary generation method. The role for CCUS in this pathway is greater than the other pathways with over 170 $MtCO_2e$ being captured and stored in 2050. Because hydrogen is the main energy source for heating and transport, electricity demand and therefore generation is lower than the other pathways at around 340 TWh (around the same level as today).

7.2.1 To Blend or not to Blend?

As the UK contemplates such a bold future for hydrogen in domestic heating, a key question arises whether the hydrogen should be blended in with the existing natural gas supply or whether the system should be converted to the use of carbon-free hydrogen.

The notion of blending relates to the proposition that hydrogen might be mixed with natural gas prior to distribution. A typical proportion might be 20% hydrogen by volume. The blended gas proposition has the following benefits:

- The calorific value and combustion properties of the blended gas can lie within limits established for natural gas. As such, there is no need to undertake conversions at downstream combustion points—such as domestic boilers.
- All substitution of natural gas with hydrogen represents a significant, but admittedly far from complete, move towards energy decarbonisation and one that can be achieved at relatively modest cost.

The principle shortcoming is that the scale of the potential benefit is capped by the limits placed on the hydrogen proportion in the blend. As such, decarbonisation of the gas system is not possible via this approach.

Ideas focussing on the complete local replacement of natural gas with hydrogen are bolder and more expensive. In such case, end-user combustion system conversions are required.

Both blended and unblended approaches will introduce hydrogen molecules into the gas distribution system. Concerns have been raised that the hydrogen might damage the pipes of the gas distribution grid. The embrittlement of steels by hydrogen is a well-known phenomenon. Whether such effects actually pose a significant risk to the legacy UK, pipeline infrastructure is a somewhat separate question. Indeed for decades prior to the conversion to natural gas in the late 1960s, the UK operated pipe networks to distribute "town gas" a syngas comprising primarily hydrogen and carbon monoxide. That activity was done successfully with few pipeline materials concerns. Since then, the industrial use of hydrogen in steel pipework has been extensive, for example in the steel industry. In seeking to assess the potential pipework damage caused by hydrogen, one must be careful to consider the damage that might be caused by minor impurities. This would appear to be particularly important when assessing historical damage to industrial hydrogen infrastructures as deployed in petrochemical plants, for example.

Readers interested in the issues relating to the possibility of hydrogen embrittlement in pipeline steels are recommended to consider the work of Hardie et al. [8]. For plastic pipes, there are also concerns that hydrogen might cause reduced ductility or a worsening of other mechanical properties under certain operating conditions [9]. In the UK, Northern Gas Networks is engaged in a major study to assess gas pipelines with a view to hydrogen transmission and distribution.

In this chapter, we will look in more detail at two UK projects being proposed to deploy hydrogen at scale within existing regional gas distribution systems. The first of these is the "h21" proposal originally associated with the Leeds Gateway organisation and aiming for the distribution of 100% unblended hydrogen.

7.3   Leeds Gateway "H21" Proposition

The Leeds Gateway h21 proposal is particularly bold in its ambition. The intention is to start with a focus on a single region in the north-east of England including the cities of Leeds and Kingston upon Hull. The area lies adjacent to a well-established chemical engineering cluster and offshore assets associated with natural gas extraction.

The results of a substantial initial study were published in 2016 [10]. From that report, the scale of the ambition first became clear. For the limited region described, it is reported that the existing natural gas demand is substantial, equivalent to a continuous average of: 678 MW with a peak hour demand of 3180 MW and a peak day average demand of 2067 MW [10]. To meet this demand with hydrogen, the 2016 report proposed to deploy four steam methane reformers on Teesside collectively supplying 1025 MW$_{HHV}$ of hydrogen (equivalent to 305,000 standard cubic metres per hour). The 2016 vision further proposed that the system will be integrated, see Fig. 7.6, with 140 bar carbon capture and storage at 90% efficiency and inter-seasonal hydrogen storage in local salt caverns providing confidence that peaks in demand can be met, as needed.

The 2016 analysis observed that a key part of the initial costs will be the individual end-user property conversions required when shifting to 100% hydrogen. The overall costs of conversion are estimated to be £3,078 per domestic property, and it is noted that this is similar to the £3,500 costs seen in the 2011–2012 Isle of Man conversion programme where properties on the island were converted from town gas to natural gas [10].

After the 2016 activity, three companies (Cadent, Equinor and Northern Gas Networks) came together as H21 North of England (H21 NoE) to prepare a detailed engineering concept for the phased roll-out across the UK of a pipeline natural gas to hydrogen transition. The aim was to consider a significant scaling up of the original Leeds Gateway concept of 2016. The planned phased roll-out would comprise a first phase at least 10 times larger than the 2016 vision. The full implementation would be a scheme 50 times larger than the original 2016 ideas. On 23 November 2018 at

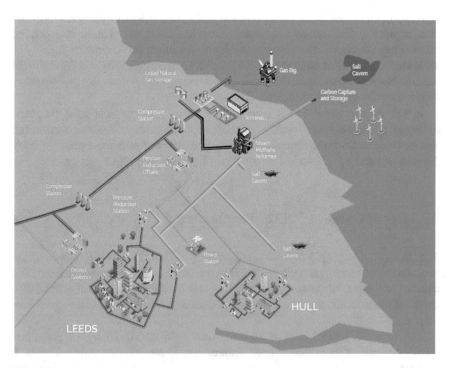

Fig. 7.6 h21 proposal for north-east England, here illustrating the fullest plans with hydrogen storage in salt caverns, as envisaged in 2016. *Source* Leeds City Gateway h21 Report 2016 [10]

the Institution of Mechanical Engineers in London, H21-NOE launched their results to a packed house in the form of a hefty 537-page report [11]. It certainly presents a bold vision. Key considerations include:

- The end-point challenge is the conversion of supply to 16 million homes in the UK.
- Industrial-scale carbon capture and storage is attracting growing stakeholder interest.
- The first phase would occur between 2028 and 2034 with the aim of reducing UK CO_2 emissions by 12.5 million tonnes per annum.
- This first phase will decarbonise 14% of UK heat by 2034 at a cost of GBP 22.7Bn.
- The hydrogen will be produced from natural gas in a new 12.15GW facility.
- This hydrogen will serve 3.7 million consumers representing 17% of the UK total.
- Inter-seasonal hydrogen storage in the form of ammonia in natural salt caverns appears to be the most technically attractive option.
- The aim is for 8TWh of inter-seasonal storage.
- Costs would be socialised across all gas customers (NG and H_2) and not be charged to just the newly converted hydrogen customers.
- The costs would represent a 7% increase in consumers' bills.

The engineering proposed by H21-NoE is certainly bold and inspiring. It truly does set out a means by which material decarbonisation progress might be made at a cost comparable to business as usual infrastructure replacement (see, e.g. the proposed renewal and expansion of the UK nuclear power plant fleet). Broadly, the H21-NOE proposition is also powerful in an entirely different way. It is politically compelling (c.f. discussion on p. 80). The ideas are clearly substantial representing significant capital investment and opportunities for employment with costs to be spread across the whole country. The politically and socially important reality is that the ideas relate very strongly to the north of England. Simply put, this represents an opportunity for northern leadership and for regional industrial regeneration. This could not come at a more important time for a region that has experienced decades of deindustrialisation as a result of globalisation and in a country that is questioning the national relationship to the global economy as never before. As Britain approached the end of the first quarter of the twenty-first century, she needs bold, almost heroic, projects. The people of the UK generally, and in some regions most particularly, deserve a renewed sense of pride. The H21-NOE vision manages to combine the boldest of twenty-first century energy engineering with industrial heritage and notions of regional and national identity. It further has the attribute of putting substantial investment into parts of the UK that have seen social deprivation and which have expressed political dissatisfaction. This is an opportunity to provide substantial investment into what is arguably part of "left-behind Britain".

7.4   HyNET Project in the North-West of England

Across the Pennines in the north-west of England, another hydrogen conversion vision is taking shape—the HyNET project. The plan here is to deploy hydrogen as a blend with natural gas. The HyNET plan can be regarded as complementary to, rather competitive with, the h21 proposals discussed earlier. The reason is that HyNET is far more closely aligned with today's industrial and regulatory realities. In addition, some of the infrastructures proposed could be very helpful to the later H21-NOE ambition. In particular a near-term development of HyNET will boost skills and experience that would later be hugely helpful to an h21 roll-out.

Some selected key attributes of the HyNET philosophy may be summarised from [12, p. 11].

- "To be the lowest cost approach to heat decarbonisation compared to alternatives"
- "To provide a material level of CO_2 abatement that is deliverable"
- "To match the project's risk profiles with Business as Usual (BAU) risk profiles"
- "To be a significant, but 'no regrets' step forward, which can be supported by policy-makers".

Launched publicly in May 2018, the HyNET project is a £900 million vision for a major hydrogen and CCS project in the north-west of England (Fig. 7.7). The project is supported by the regional gas distribution company Cadent. Cadent is responsible for the delivery of piped natural gas to domestic and industrial consumers in four regions of the UK: the West Midlands of England, North-West England, the East of England and North London.

The aim of the HyNET project is to supply low-cost hydrogen for heat applications coupled with low-cost CCS. The plan is to achieve 1MT of CO_2 captured and stored

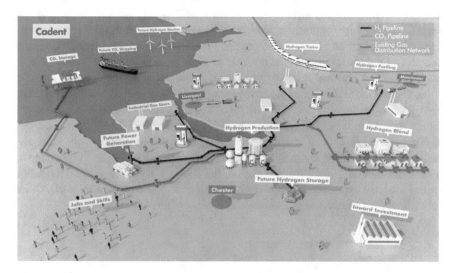

Fig. 7.7 HyNET project proposed for the north-west of England, with thanks to Cadent [12]

annually. In the summer of 2018, the HyNET project was assembling the funding that will be needed to permit the roll-out of the project.

The initial aim for HyNET is to decarbonise 10 heavy industrial operations and to deliver hydrogen gas to 2 million homes via a 20% blend with natural gas. At that level of dilution, no adjustment to downstream equipments will be required and combustion parameters will be within existing technical limits for end-user appliances.

The HyNET plan envisages that the CCS will be close to shore, which represents a significant economic advantage. In the north-west of England, it is convenient that there are local depleting offshore fossil fuel extraction assets nearing the end of their production life. Furthermore, one of the UK's two big fertiliser plants is nearby and that also has the potential to be very useful because that facility already produces and captures CO_2. In principle, the fertiliser plant technology could be linked to the HyNET CCS technology. Additionally, it could be used to test and commission the HyNET CCS technology, even before the CO_2 from the HyNET process starts to flow.

The HyNET project team estimates that the hydrogen produced, when used for heat, will be available at only 10% of competitor cost (per Tonne of CO_2 saved). Using HyNET, hydrogen to produce domestic heat has a second benefit in that it avoids the need to scrap all the existing natural gas boilers. Indeed even a conversion programme is not needed, given the plans to supply hydrogen as a 20% blend with natural gas. The vision of the HyNET team seeks to minimise the stranding of existing assets including assets (such as boilers) belonging to domestic end-users. It is in this way that the HyNET vision aims to be ultra-low cost.

It should be noted that alongside the HyNET project there is a separate project at Keele University known as HyDeploy looking at the safety case of deploying hydrogen blends in the UK. In 2019, this project is making good progress and is navigating regulatory requirements well.

The possible future linkages of HyNET to other wider hydrogen activities are shown in Fig. 7.8.

Arguably, the hydrogen path to decarbonisation (either as a blend or as 100% hydrogen) is fundamentally not a technical challenge, as all the necessary technologies already exist. The obstacles to deployment all lie in the domains of economics and policy. Most especially there are a series of "chain risks". We here observe that these are similar to the risks facing any network industry, but in this case the issues are especially deep, given the high capital intensity of some of the technologies. In essence, the plan only delivers compelling value once all its elements are in place. The deployment phases for several of the key technologies are not quick, and hence, funding during deployment can represent a daunting challenge to project proponents. The case for state intervention, at the level of industrial strategy, is strong.

In presenting the H21-NOE and HyNET projects, we have emphasised two different hydrogen distribution approaches. There is, however, another technical difference between the two projects. In the north-western (HyNET) plan, the aim is to use autothermal reforming, whereas the 2016 Teesside/Leeds Gateway (h21) plans make use of steam methane reforming (SMR). Autothermal reforming was introduced in Chap. 3. To recap: it comprises partial oxidation followed by catalytic reforming.

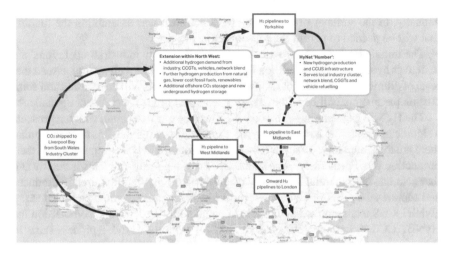

Fig. 7.8 Possible extensions to the HyNET project, with thanks to Cadent [12]

Recalling Eq. 3.3, for a natural gas feedstock, the relevant equation is:

$$2CH_4 + O_2 + CO_2 \rightarrow 3H_2 + 3CO + H_2O$$

The emphasis on useful carbon monoxide production (for chemical applications) is reminiscent of the activities in the USA along the Gulf of Mexico (see Chap. 6).

A key issue shaping the prospects for roll-out of natural gas sourced hydrogen, beyond initial relatively small-scale pilot projects, would appear to be the development and demonstration of low-cost CCS solutions. Key regions of the UK are superbly placed to pioneer such innovations in Europe.

7.5   The Wider UK Context

These two examples above focus on the role that hydrogen can play in helping to decarbonise UK heat. As explained in earlier chapters, heat and mobility are the two key sectors requiring action if the UK is to hit its ambitions 2050 decarbonisation targets. As we approach 2020, it is clear that the easiest decarbonisation tasks have now been done and most of the action has been in the domain of electricity. But the bulk (approximately 65%) of the task still lies ahead and this requires a significant new push of effort and ambition. H21-NOE and HyNET represent exciting and ambitious projects in the heat area.

It is important to stress that if the UK does make a bold move into hydrogen for heat, then there will be beneficial knock-on consequences for hydrogen mobility. The UK government has an eye to such hydrogen-based mobility with the Clean Growth Strategy reporting: "*The Government has provided £4.8 million through*

the Hydrogen for Transport Advancement Programme to create a network of 12
hydrogen refuelling stations, and £2 million through the Fuel Cell Electric Vehicle
Fleet Support Scheme to increase uptake of hydrogen fuel cell cars and vans in the
public and private sector. A new £23 million fund was recently announced to boost the
creation of hydrogen fuel infrastructure and encourage roll-out of hydrogen vehicles"
[7, p. 88].

7.6   Ammonia and the World

The 2018 H21-NoE report is noteworthy for the emphasis given to ammonia as a
hydrogen storage method [11]. The report notes that with a boiling temperature at
atmospheric pressure of 240 K (-33 °C), ammonia presents only modest challenges
at a level similar to liquefied petroleum gas (LPG). The ammonia production industry
is well-established and well linked to issues concerning hydrogen. Ammonification
of nitrogen with hydrogen is already extremely well-established (see Chap. 1). The
H21-NoE report of 2018 considers the technical and economic issues of ammonia-
based storage in some depth [10] and further suggests that ammonia might not just
meet the needs of UK inter-seasonal storage but also eventually permit a global and
interlinked hydrogen economy with hydrogen cargoes sailing between markets as
ammonia in fully-refrigerated ships not unlike the LNG tankers of today. In this way,
a British industrial innovation (the transition from natural gas to hydrogen supply)
has the potential not just to act to reduce UK emissions but to help wider global moves
to decarbonisation. The potential for the UK's large international oil companies to
be enthusiastic about such a shift away from the retail sale of fossil fuels will be
discussed in Chap. 10.

Acknowledgements In preparing this chapter, we are most grateful to advice and information
provided by Damien Hawke, Alaric Marsden, Dave Robson, Dan Sadler, Simon Schaeffer and
others. It is important to stress that those that have provided assistance to this chapter do not
necessarily agree with what is said. All responsibility for what follows lies with the authors.

References

1. The Climate Change Act. 2008. January 4, 2019. Available from: https://www.legislation.gov.
   uk/ukpga/2008/27/contents.
2. Maslin, M. 2014. *Climate: A very short introduction*, 187. London: Oxford University Press.
3. Marsden, A. 2018. *Gas as a key enabler of flexible electricity generation*. The role of Gas
   in the energy transition 2018. Available from: https://www.eprg.group.cam.ac.uk/wp-content/
   uploads/2018/05/Alaric-Marsden.pdf.
4. Goodall, C. 2016. *Better average outputs will mean UK wind will frequently meet entire
   national electricity needs*. January 14, 2019. Available from: https://www.carboncommentary.
   com/blog/2016/9/5/better-average-outputs-will-mean-uk-wind-will-frequently-meet-entire-
   national-electricity-needs.

5. Smith, B. 2019. *BEIS to look again at smart meters programme costs amid overspend warnings.* January 14, 2019. Available from: https://www.civilserviceworld.com/articles/news/beis-look-again-smart-meters-programme-costs-amid-overspend-warnings.
6. Grant Wilson, I.A. 2016. Energy data visualization requires additional approaches to continue to be relevant in a world with greater low-carbon generation. *Frontiers in Energy Research* 4 (33).
7. The Clean Growth Strategy Leading the way to a low carbon future. 2017. January 16, 2018. Available from: https://assets.publishing.service.gov.uk/government/uploads/system/uploads/attachment_data/file/700496/clean-growth-strategy-correction-april-2018.pdf.
8. Hardie, D., E.A. Charles, and A.H. Lopez. 2006. Hydrogen embrittlement of high strength pipeline steels. *Corrosion Science* 48 (12): 4378–4385.
9. McWilliams, A. 2018. *The global hydrogen economy: Technologies and opportunities through 2022.* BCC Research.
10. H21. 2016. January 14, 2019. see specifically section 3.6. Available from: https://www.northerngasnetworks.co.uk/wp-content/uploads/2017/04/H21-Report-Interactive-PDF-July-2016.compressed.pdf.
11. Sadler, D., et al. 2019. *H21 North of England,* January 27, 2019. Available from: https://northerngasnetworks.co.uk/h21-noe/H21-NoE-23Nov18-v1.0.pdf.
12. Cadent. *HyNET Northwest from Vision to Reality.* 2018. January 27, 2019. Available from: https://hynet.co.uk/app/uploads/2018/05/14368_CADENT_PROJECT_REPORT_AMENDED_v22105.pdf.

Chapter 8
Small-Scale Local Hydrogen Production

In Chap. 2, we considered the emergence of a commercially viable and politically sustainable future for mobility emerging from the mature hydrogen businesses and infrastructures existing today. Such a future has the potential to emerge in regions such as the US Gulf Coast and the Benelux region of Europe where there is already well-established hydrogen infrastructure (see Chap. 6). We shall term these established industrial infrastructures (and their successors) "Big Hydrogen". In this chapter, however, we consider an alternative possible pathway one that does not require established petrochemical facilities and hydrogen pipelines. We here consider a proposition with the potential to emerge in innumerable cities in many developed economies. The vision described here merely imagines an urban location with an established natural gas distribution infrastructure and a standard electricity supply. We further posit that the ideas proposed in this chapter will find most acceptances in cities with legacy transport and air quality problems and in countries with sincere concern for global climate change.

8.1 The Proposition

The goal outlined here is to facilitate a shift for local vehicles from petroleum-based vehicles to hydrogen-fuelled fuel cell electric vehicles (FCEVs). The closest competing paradigm for low emission mobility in urban contexts is the use of electrically charged battery electric vehicles. That counter-scenario presents the following potential difficulties:

- The practical difficulties of apartment dwellers to charge their vehicles from home
- The challenge and cost of deploying a public charging infrastructure
- The time-varying stresses caused to the urban electricity distribution system.

The challenges are summarised in Table 8.1.

© Springer Nature Switzerland AG 2020
W. J. Nuttall and A. T. Bakenne, *Fossil Fuel Hydrogen*,
https://doi.org/10.1007/978-3-030-30908-4_8

Table 8.1 Competition between battery electric vehicle and hydrogen fuel cell vehicle power systems in the Small Hydrogen Scenario

Issue	Battery electric vehicles	H$_2$ fuel cell vehicle
GHG emissions:		
At tailpipe	Zero	Zero
Other system "well to wheels" emissions	Typically significant today, depends on electricity system emissions especially the role of unabated fossil fuel combustion and whether the electricity used can genuinely be identified as renewable or nuclear	Typically significant today, but potentially low in the future. The prospects for low-GHG hydrogen production are described in this book
Urban air quality	Improved—displaces existing polluting vehicles	Improved—displaces existing polluting vehicles
User groups	BEVs are favoured by fleet users for smaller vehicles (e.g. taxis). BEVs also gain significant market share with private users	FCEVs feature first with fleet users and tend to be oriented to larger vehicles initially (e.g. buses and trucks). FCEVs lag BEVs with private users but this could change rapidly as H$_2$ infrastructure improves
Natural gas infrastructure	Irrelevant	Necessary for 'Mature Hydrogen' approach. Irrelevant for 'Green Hydrogen' approach
Electricity supply	Very important, but full impacts still unclear … there is even a risk of worsening, rather than ameliorating, electricity system stresses	Irrelevant
Harsh cold winters	Vehicle passenger space heating is a problem	Few problems here (many fuel cell designs run at high temperature)
Regional manufacturing industry	Irrelevant	Relevant to CO/CO$_2$ sale/disposal as part of CCU

Of course, private FCEV and BEV mobility are not the only possible urban futures. Enhanced public transport might reduce the need for private transport. Socio-technical change might additionally yield new modes of living and working requiring less personal mobility. With that said, one can however posit a continuing role for cars (if only as taxis) in the future of urban mobility. In that context, the deployment of an urban hydrogen production infrastructure becomes particularly interesting.

We have suggested previously (Chap. 2) that the first deployments of hydrogen vehicles will be for larger vehicles especially those associated with longer-range use. Long-distance road haulage would be a clear example. A follow-on example would be metropolitan bus services. After these easiest and earliest examples, one might imagine distinct vehicles fleets dictated by a combination of business choice and local licensing conditions. A clear example of such a situation would be a licensed taxi fleet in a major metropolitan area. It is already the case that many cities already constrain the types of vehicle eligible for a licence. We shall keep the taxi fleet example in mind as we posit a transition to a more environmentally responsible Small Hydrogen future.

We have already suggested in this book that many of those arguing for Green Hydrogen futures on the grounds of perceived environmental benefits have tended to neglect the reality of Big Hydrogen infrastructures and the possibilities for beneficial innovation within such systems. As noted earlier (Chap. 3), SMR CO_2 emissions are associated with two somewhat separate aspects of the process: reagent heating and the chemical process itself including the water-gas-shift conversion of CO. For deeper insight into these issues we recommend the work of Lindsay et al. [1]. As discussed in Chap. 6, the largest Big Hydrogen infrastructures, such as the Air Products pipeline-connected assets along the US Gulf Coast, provide ample technical opportunities to greatly reduce both process and flue gas emissions associated with fossil-fuel-based hydrogen production.

As a deliberate counterpoint to "Big Hydrogen", the vision presented in this chapter will be called "Small Hydrogen".

8.2 Small Scale Alternatives

Arguably, we are far from the first to posit the benefits of Small Hydrogen. For example, in the UK, the Research Councils have funded a large-scale collaborative inter-university research programme. The "Delivery of Sustainable Hydrogen (DOSH2)" research programme which formed part of the Research Councils UK "Supergen" programme. The research focussed on the potential for local hydrogen production as a widely distributed activity involving mostly small-scale hydrogen generation technologies. We suggest, however, that we are relatively unusual in restricting our core attention to hydrogen produced from natural gas.

In contrast to the Big Hydrogen futures introduced in Chap. 6, the pre-requisite for the emergence of the Small Hydrogen alternative is not the existence of a well-established hydrogen pipeline infrastructure and the local existence of advanced chemical processing industries, rather Small Hydrogen is favoured by a highly urbanised area with the following attributes:

- A local political desire to do, and be seen to be doing, something about climate change
- A multi-stakeholder desire for improved urban air quality
- A large number of taxis, or other fleet users susceptible to special local regulatory constraints
- Well-established legacy natural gas supply infrastructures
- Stressed electricity supply infrastructures
- Seasonal energy system stress
- Less than 100 km from industrial/manufacturing operations requiring CO_2.

By way of example, such location might be Manhattan Island in New York City or the inner London congestion charged zone. In these contexts, the primary motivations for change could come from the first two ideas above, namely: climate change mitigation and urban air quality improvement. One can already see that Business-as-Usual scenarios will include ever more stringent tailpipe emissions limits for key communities going forward. One clear extension of policy would be the shift to zero-emission taxi operations. Indeed, taxi manufacturers are already preparing for such realities, albeit based on Plug-In Hybrid Electric Vehicle technology, see, e.g. [2]. Looking further ahead, there is indeed the prospect of a hydrogen-fuelled London Taxi, see e.g. [3].

As we consider the prospects for urban mobility, for example, taxi fleet operations, two competing technologies battery-based electric vehicles (including both BEVs and PHEVs) and hydrogen fuel cell vehicles emerge as direct competitors. As things stand, the BEV and PHEV way ahead appears to have the greatest traction, but might this change with time and experience?

8.3 Relative Merits of Local Hydrogen from Electricity and Natural Gas

Let us imagine a future scenario in the UK in the year 2035.

In 2035, the UK is leading the world in rolling out a genuine hydrogen economy based on natural gas and upstream CCS. By the mid-2030 s, this transformation is rolling south from the north of England. It all started with heat, but increasingly the easy availability of abundant hydrogen is making in-roads into mobility. The first mobility applications were buses and trucks, but now we see the first affordable FCEV cars—they appeal especially to a large chunk of those drivers who have thus far hung on to traditional internal combustion vehicles and fossil fuel hybrids. They tend to be middle-income people usually with no off-street parking and whose driving habits justify their "range anxiety". The new hydrogen cars (FCEV) meet their needs, and in the cities of the north, you can see hydrogen cars filling up quickly at nearly all the filling stations. Those stations have the necessary gas clean-up kit and the compressors needed for high-quality hydrogen supply. They access hydrogen from the transitioned hydrogen gas transmission and distribution network (as introduced in Chap. 7).

The prospect of a phased decarbonisation of heat, as described in Chap. 7, illustrates a phased conversion from natural gas pipeline gas to hydrogen. Hence increasingly, there will be parts of the country with readily accessible pipeline hydrogen supply. We suggest that vehicle filling stations belonging to the international oil companies could relatively easily install the clean-up kit sufficient for that hydrogen to be sold for FCEV use. For the reasons described earlier in this chapter, one could imagine a rapid take up of hydrogen-fuelled FCEV cars in that part of the country. Ordinary motorists are not as careful in planning their refuelling as fleet operators and it is to be expected that these motorists will need to refill their cars while driving in other, still to be converted, parts of the country.

In Southampton on the South Coast of England in 2035, there is still no hydrogen infrastructure, and even in the north of England, there are towns that have been missed for various reasons. These towns in the north and cities in the south are still receiving natural gas by pipeline—not hydrogen. How can hydrogen car drivers from the north refill in the south of England for their journey home? How can those that live in the south buy a decent new car that doesn't need electricity charging, now that petroleum vehicles (including HEVs) are truly fading out and are becoming harder to buy? It remains the case that those that live in flats, or that drive long distances, are not attracted to battery electric vehicles.

So, how are demands for clean hydrogen supply in the south of England or for northern towns not yet on the hydrogen network being met? How does supply evolve and move as the network changes over?

Solution: a low cost, semi-portable, "pop-up" hydrogen supply point. Ideally, it should fit in one (or at most two) standard ISO containers. The system should be relocatable as the frontier of hydrogen availability moves across the country.

Fundamentally, there would appear to be two ways to meet hydrogen demand subject to these constraints. The more established proposition would be to make use of available electricity (preferably low-carbon electricity) and use local electrolysis of water to produce high-quality hydrogen. The alternative, of interest to us, is to make use of a long-established natural gas connection and use SMR technology to make hydrogen, preferably with associated CCS or CCU provision.

In summary, we envisage:

- Modular units, perhaps based on a standard intermodal (or 'ISO') shipping container
- On-site production of hydrogen from natural gas
- On-site capture of carbon dioxide.

Please note that for the example locations of New York City and London, there is already a high degree of official understanding of the issues across key sectors of the city's economy. There is much existing awareness in New York City [4], and we posit that the Manhattan Island context is fertile for the necessary policy push.

There is one key potential innovation that is potentially highly relevant to the Small Hydrogen vision—carbon capture utilisation and storage, as discussed in Chap. 5. It should be stressed that for the Mature Hydrogen business the CCS prospects are already good. There is much relevant technical understanding in the domain of Big Hydrogen including that associated with Enhanced Oil Recovery (EOR). Follow-on research in this area can be expected to consider synergies with CCS in greater depth. In contrast, the natural gas-based Small Hydrogen pathway appears less developed as a concept. Such work as has been done on Small Hydrogen ideas has tended to focus on electrolysis-based methods especially those related in some way to renewable electricity.

In future work, we aim to compare (cost, environmental impact and safety) the potential for small-scale natural gas-based hydrogen production with the alternative

Fig. 8.1 The University of Texas at Austin SMR-based Hydrogen production, storage and Filling station. This includes many of the elements that would be expected in an example of a Small Hydrogen installation. *Source* Authors (WJN) with thanks to UT-Austin

hydrogen production paradigm of advanced electrolysis-based production. In particular, we intend to focus on similar small scales of hydrogen production for urban use. In this way, we hope to gain a direct insight into the strengths and weaknesses of the two potential Small Hydrogen production methods. In looking at solutions that might be up-scaled from small scale pilots, we note that the comparison would be timely as well as being tractable as data relating to both alternatives should be relatively accessible given the existing moves to implementation (see, e.g. Fig. 8.1).

We further note that ITM Power in the UK has recently developed a rapid response electrolysis capability explicitly designed to match to opportunities of very low-cost power expected in electricity markets with high penetrations of intermittent renewable generation. Indeed, ITM Power has already sold to Shell a 10 MW hydrogen production facility designed to fit on a standard skid easily delivered and deployed to site [5].

We observe that Small Hydrogen might also be coupled to on-site hydrogen storage and on-site electricity generation (via robust high-efficiency stationary fuel cells). With such capability, the installation could, in addition to sourcing valuable hydrogen, also respond sufficiently quickly to provide electricity grid balancing services which might be expected in future urban power balance scenarios to be well-remunerated. In this way, Small Hydrogen might actually link well to both the existing natural gas infrastructure and the evolving electricity infrastructure.

References

1. Lindsay, I., C. Lowe, S. Reddy, M. Bhakta, and S. Balkenende. 2009. Designing a climate friendly hydrogen plant. *Energy Procedia* 1:4095–4102.
2. The Future Zero-Emission London Taxi. January 19, 2019. Available from: https://www.levc.com/corporate/news/the-future-zero-emissions-london-taxi/.
3. Case Study: Hydrogen Fuel Cell Taxi. January 19, 2019. Available from: https://www.lotuscars.com/engineering/case-study-hydrogen-fuel-cell-taxi.
4. Pasion, C., M. Amar, and M. Delaney. 2016. *Inventory of New York City Greenhouse Gas Emissions November 2014*, in *Mayor's Office of Long-Term Planning and Sustainability*. New York.
5. ITM Power. 10 MW Refinery Hydrogen Project With Shell, September 1, 2017. Available at: http://www.itm-power.com/news-item/10mw-refinery-hydrogen-project-with-shell.

Chapter 9
Hydrogen Cryomagnetics—A Physics-Based Innovation

Thus far in this book, we have generally concerned ourselves with the potential role of hydrogen gas as an energy carrier. In Chap. 2, however, we pointed to the possible use of cryogenic liquid hydrogen in the context of vehicle refuelling systems. In this chapter, we focus specifically on the attributes of liquid hydrogen, the role it might play in the emerging hydrogen economy and some special benefits that arise as a consequence of the very low temperatures involved.

One of the authors of this book (WJN) came to consider hydrogen futures as a consequence of having spent much effort, with others, examining the long-term future of helium [2]. That journey went as follows… Helium is an important and rather special substance that is posited to play a major role in our high-technology future. Much of the global use of helium relates to its use as a cryogen cooling superconducting wires, for example, in high field magnets. Helium cooled superconducting magnets have a range of high technology uses including in magnetic resonance imaging medical scanners used to image soft tissues for medical diagnosis. In the transport sector, superconductivity has long been associated with the prospect of magnetically levitated high-speed trains. More mundanely superconductivity could make possible lightweight very high torque electric motors. Generally, superconductivity is an enabler of high technology innovation.

It seems probable that global interest in superconducting magnets will increase in the coming decades, placing ever greater demands on helium supply. There is a certain irony therefore in the reality that this substance of potentially great importance to the late twenty-first century is a by-product of the fossil fuel economy of the twentieth century.

Historically, helium has been a by-product of the natural gas industry, and while the global helium industry is a roughly one billion dollar business, it sits within the one trillion dollar natural gas business. It is therefore unsurprising that the natural

Part of this chapter has been published previously in the Institute of Physics report *Next Steps for Hydrogen—physics, technology and the future* by William J. Nuttall, Bartek A. Glowacki and Satheesh Krishnamurthy [1]. As such, this chapter includes material © Copyright the Institute of Physics, 2016. Reproduced under permission.

gas industry does not always make the best decisions concerning the sustainability of helium stocks. Far too much precious helium is simply vented to the atmosphere by the upstream natural gas industry each year. The challenges facing helium going forward have been summarised by Nuttall et al. [3].

The sense that helium presented a set of worrying issues concerning: on the one hand, supply-side wastage and connection to a fossil fuel industry under pressure; and on the other, anticipated growth in demand to meet increasing high technology needs, led us to the observation that an alternative might be needed, and the alternative that we identified was liquid hydrogen [4]. It became clear that liquid hydrogen at 20 K might have something to contribute to the future of superconductivity displacing demand for scarce liquid helium. In this way, the notion of "hydrogen cryomagnetics" emerged [5].

9.1 Hydrogen Cryomagnetics—An Introduction

At its heart, the idea of hydrogen cryomagnetics is that liquid hydrogen can simultaneously play two valuable roles: as a fuel and as a coolant. There is merit in bringing together the ability to generate electricity in a hydrogen fuel cell and the ability to cryogenically cool the windings of a superconducting magnet, for example, in a high torque electric motor. These attributes taken together could permit a whole range of disruptive innovations especially in low carbon mobility.

The key cryogenic attributes of liquid hydrogen are revealed by the hydrogen phase diagram, Fig. 9.1.

For hydrogen cryomagnetics, the key beneficial attribute is that at atmospheric pressure, hydrogen liquid boils at 20 K (−253 °C) and perhaps slightly more problematically solid hydrogen forms below 14 K (−259 °C). In essence, these realities allow for relatively straightforward cryogenic cooling of equipment to temperatures at which modern superconductors can operate. The key attribute of such superconductors is that once at a sufficiently low operating temperature, they can pass an electrical current with effectively zero electrical resistance. This makes possible far more efficient electromagnets, as used in electric motors, for example.

While we look ahead to the possibility of hydrogen cryomagnetics, it remains the case today almost all commercial applications of superconductivity rely on the use of helium in some way. Helium has the lowest boiling temperature of any known material (4.2 K at atmospheric pressure), but as noted previously, its continuing availability requires a continuing fossil fuel economy.

We mentioned earlier the role played by liquid helium in magnetic resonance imaging (MRI). Increasingly, MRI systems are using closed-circuit mechanical cryocoolers to achieve the necessary low temperatures, and hence, helium wastage is minimal in the clinical environment. The MRI industry, however, continues to require very large amounts of liquid helium in the MRI scanner manufacturing process at the factory. Equipment testing and cooling for delivery are the main processes demanding liquid helium.

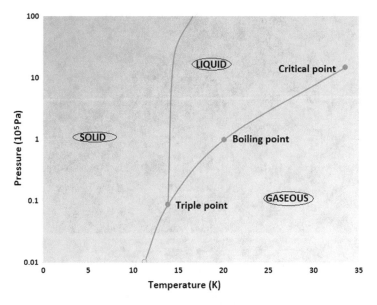

Fig. 9.1 The phase diagram of hydrogen adapted from [6]

The vision of hydrogen cryomagnetics has at its heart the use of liquid hydrogen at its atmospheric pressure equilibrium temperature of 20 K, combined with the use of modern higher temperature superconductors (HTS), to achieve magnetic services and solutions free from the risks of escalating helium prices. As such, hydrogen cryomagnetics enters a competition with mechanical recirculating cryo-coolers (with their associated energy costs) or with industrial substitution away from superconductivity altogether.

9.2 Pathways to the Hydrogen Economy

In 2006, Joan Ogden looked ahead to the emergence of a hydrogen economy. In Fig. 9.2, we reproduce a schematic diagram from her paper to make the point that a cryogenic tanker truck supply chain for liquid hydrogen is one of the easiest early modes for hydrogen distribution (top image). This method of distribution is attractive when the quantities required remain relatively small. Once very large quantities are required, then pipeline distribution becomes preferable and this links to several of the ideas considered previously in Chap. 6. The thinking illustrated in Fig. 9.2 is independent of the possible merits of hydrogen cryomagnetics, which has the potential to give even greater impetus to cryogenic approaches.

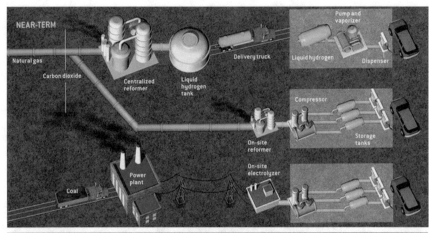

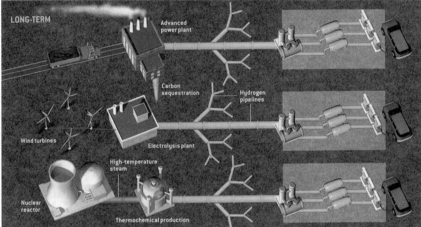

Fig. 9.2 Hydrogen Supply Systems—near-term (top) and long-term (bottom). *Source* Artwork by Don Foley (rights holder). As originally published in Scientific American [7]

9.3 The Attributes of Liquid Hydrogen

To exist as a liquid, H_2 must be cooled below hydrogen's critical point of 33 K (-240 °C), see Fig. 9.1. The critical point is a set of ambient parameters (including temperature) above which the fluid phase shows no clear distinction between gas and liquid. However, for hydrogen to be in a full liquid state without vaporising at atmospheric pressure, it needs to be cooled to 20.3 K. As noted earlier, the melting point of hydrogen is relatively close by at 14.0 K (the triple point of hydrogen is at 13.8 K 7.04 kPa), but this concern can be avoided relatively easily.

Liquid hydrogen has a set of important attributes. It has low viscosity and has a low molecular weight. It has high thermal conductivity and a very high heat of vaporisation. These attributes favour the use of hydrogen as a cryogenic coolant in comparison with other low-temperature cryogenic liquids, such as helium and neon. The future price and availability of helium for cryogenic applications have been an issue of growing concern in recent years [2].

The use of hydrogen cryogenics does not necessarily imply the elimination of helium from the cryogenic cooling system. Indeed, there are benefits in ensuring the proximate cooling of components is achieved with dense helium gas at 20 K. That is, hydrogen cryomagnetics can be combined with helium in an indirect closed-loop cooling scheme. In such a scheme, 20 K gaseous helium is in thermal equilibrium with liquid hydrogen. The cooling of critical components is achieved using 20 K helium (permitting high system pressures, ensuring high heat transfer rates and protecting against safety risks and solid-phase blockages [3]). An excellent example of the superior efficiency of the liquid hydrogen indirect cooling over the liquid helium and liquid neon cooling has been demonstrated by McDonald et al. in a 15 T pulsed non-superconducting copper solenoid magnet at the CERN Laboratory [8].

The liquefaction of cryogenic gases requires significant energy. When considering the joint attributes of liquid hydrogen as an energy carrier and as a coolant, it is important to understand what proportion of the energy embodied in the original gaseous hydrogen would be required to cool that hydrogen into the liquid state. In large-scale liquid hydrogen production (e.g. quantities of ~10,000 kg/h), the process of liquefaction requires energy equivalent to roughly 30% of the higher heating value (HHV) of the hydrogen. While this might appear to represent significant process inefficiency, converting hydrogen into the more energy-dense liquid state, and into a state which has cryogenic utility, can nevertheless be very valuable given the benefits in supply and use that later arise. Small-scale hydrogen liquefaction is perhaps more challenging (e.g. approaching 1 kg/h), because then, even with advanced liquefaction systems, the energy required for liquefaction is equal to the energy carried by the hydrogen itself and that logic motivates large-scale production. Indeed, such realities very often underpin arguments entirely against small-scale hydrogen liquefaction. So, notwithstanding the earlier supply chain comments linking cryogenic liquid hydrogen to a smaller hydrogen economy, it is clear that cryogenic liquid hydrogen is best suited to those visions of the hydrogen economy emphasising large-scale centralised production. In such large-scale production scenarios, tanker-truck liquid hydrogen distribution must be compared with gaseous hydrogen supply chains which may at such scales involve pipelines. In the absence of consideration of superconductivity-based magnetism, the issue at hand is to find a sweet spot where the production industry is large enough to merit hydrogen liquefaction and the distribution industry small enough to prefer liquid hydrogen over gas-based pipeline methods. One idea might be a highly centralised production point with multiple radial distribution routes supplied by LH2 tanker trucks. Bringing hydrogen cryomagnetics ideas into the mix, however, would further favour the attractiveness of the cryogenic approach.

9.4   Liquid Hydrogen Storage

One of the most efficient methods for hydrogen storage (and distribution) is to maintain it in the liquid state. The main benefit is the vastly smaller storage volumes required compared to all gaseous alternatives. However, as discussed earlier, the principle difficulty is achieving the extremely low boiling point of hydrogen (20 K). The low temperatures require the hydrogen to be kept in heat-super-insulating containers known as dewars.

We also note the potential use of adsorption-based hydrogen storage methods using porous and structured materials, such as zeolites or carbon nanoforms. These are particular examples of cryogenic approaches which can involve more readily accessible temperatures, but which also bring other more troubling considerations— such as the risk of extended timescales for vehicle fuelling. We shall not dwell on such approaches here, as the more straightforward liquid hydrogen proposition is sufficient to present the essential arguments.

The cryo-technology of liquid hydrogen storage has been improved much in recent decades primarily at the initiative of the aerospace industry. However, liquid hydrogen is rarely associated with private automobiles partly owing to its high cost. A cryo-compressed solution for the automotive industry has been presented in Chap. 2. A key concern has been if a cryogenic liquid hydrogen tank in a car was to "run dry". In that case, the tank would start to warm and require specialist intervention (presumably at a main car dealership or even more centrally to restore the system). One would not want to imagine a member of the public attempting to refill a warmed up tank on the filling station forecourt. Safety systems would surely prevent that. That said, if the cryogenic hydrogen was to warm to room temperature, the driver would still be able to drive some distance (e.g. to the main dealership). This is because the cryogenic tank would readily serve as a 350 atm hydrogen high-pressure gas tank [9]. We posit that the "run dry" warning light in a future liquid hydrogen-fuelled car is one that the driver should not ignore.

One issue affecting vehicle hydrogen use is the combined weight of the stored fuel and its container. This has been a major difficulty faced by those developing gaseous hydrogen storage systems. Generally, liquid hydrogen systems do well in such terms. Importantly, cryogenic LH2 systems surpass the long-standing target (9 wt%) established by the US Department of Energy as a goal for 2015. That policy was established on the basis of technological neutrality and was not designed to favour one technology over another. Despite the weight and energy density benefits of liquid hydrogen, the challenges of low temperatures and the requirement for heat insulator systems tend currently to make the cryogenic liquid option inappropriate for most mobile applications. However, it is important to realise that, despite the challenges, liquid hydrogen cryogenic containers for cars is an already-existing technology. As such, the bigger technological challenge has become the transfer of liquid hydrogen to the car. This is a subject of a patent by the Linde Group (see Chap. 2). Such issues, together with the lack of availability of LH2 fuelling stations, are currently holding back the development of LH2-fuelled private cars.

Noting synergies between cryogenic gases, it can be envisaged that if liquefied natural gas (LNG)-fuelled cars were to be more widely available, then this could represent a turning point for the later roll-out of LH2-fuelled cars.

Returning to hydrogen cryomagnetics, one should bear in mind that LH2 at 20 K is a very efficient coolant for medium and high-temperature superconductors. To imagine a future car with onboard cryogenics opens up a wealth of possible follow-on innovations, such as superconducting technologies in navigation, superconducting magnetic bearings and flywheel energy storage. These innovations have the potential to accelerate the market penetration of liquid hydrogen, even more so if the car will be treated as a mobile energy storage system (both power and gas).

For stationary solutions, it is important to remember that standard stainless steel dewars designed for liquid helium are identical to those intended for liquid hydrogen in terms of construction and welding materials. Therefore, the storage of hydrogen in a liquid form in large stationary helium dewars is a relatively straightforward option and is a ready solution for a decentralised hydrogen economy.

We close with the observation that hydrogen cryomagnetics has the potential to be a major technology driver in the coming decades. We cannot predict all the innovations that might arise, but: (1) new electric drive chains for road vehicles (up to and including superconducting motors in the wheels) with concomitant benefits in vehicle design and (2) low carbon "cryoplanes" for sustainable aviation are two of the very attractive ideas already emerging.

Acknowledgements We gratefully acknowledge the assistance of Professor Bartek Glowacki, Richard Clarke and Dr. Satheesh Krishnamurthy. All responsibility for errors or omissions rests with the authors.

References

1. Nuttall, W.J., B.A. Glowacki, and S. Krishnamurthy. 2016. Next steps for hydrogen. London: Institute of Physics.
2. Nuttall, W.J., R.H. Clarke, and B.A. Glowacki. 2012. The future of helium as a natural resource. In *Routledge explorations in environmental economics*, vol. 35, ed. W.J. Nuttall, R.H. Clarke, and B.A. Glowacki. London; New York: Routledge.
3. Clarke, R.H. and B.A. 2010. Glowacki. Indirect hydrogen versus helium or nitrogen cooling for fusion cryogenic and magnet systems. In *Proceedings, 23 International Cryogenic Engineering Conference and International Cryogenic Materials Conference*, Wroclaw, Poland.
4. Glowacki, B.A., W.J. Nuttall, and R.H. Clarke. 2013. Beyond the helium Conundrum. *IEEE Transactions on Applied Superconductivity* 23 (3): 0500113.
5. Glowacki, B.A., W.J. Nuttall, E. Hanley, L. Kennedy, and D. O'Flynn. 2015. Hydrogen cryomagnetics for decentralised energy management and superconductivity. *Journal of Superconductivity and Novel Magnetism* 28 (2): 561–567.
6. The EC Network of Excellence for Hydrogen Safety "HySafe". January 19, 2019. Available from: https://www.hysafe.org/download/997/BRHS_Ch1_Fundamentals-version%201_0_1.pdf.
7. Ogden, J. 2006. High hopes for hydrogen. *Scientific American* 94–101.

8. McDonald, K., and M. Iarocci. 2018. *Use of He gas cooled by liquid hydrogen with a 15-T pulsed copper solenoid magnet.*
9. Technical Assessment. 2016. *Cryo-compressed hydrogen storage for vehicular applications.* United State Department of Energy Hydrogen Program.

Chapter 10
Deep Decarbonisation—*The Role of Hydrogen*

The challenge of decarbonising our modern industrialised society is deep and complex. It is arguable that the first major response of human society was to reduce the emissions associated with electricity generation. The nuclear renaissance posited in the early 2000s has largely failed to materialise partly as a consequence of the March 2011 Fukushima Daiichi accident, but also because of rising costs and the difficulty in allocating economic risk in the years after the financial crisis of 2008/2009. Electricity emissions have, however, been seen to reduce in western countries primarily for three reasons: first the growth of renewable generation, second a shift from coal to natural gas in combustion and third a significant improvement in energy efficiency—especially relating to lighting. Broadly, there are points of optimism associated with the decarbonisation of the electricity system. In this book, we have stressed issues relating to mobility and domestic heating and considered the need for decarbonisation in these sectors too. However, even if all these initiatives are successful, carbon emissions will still be problematic and that is a consequence of the very high levels of reduction required (in excess of 80% reductions compared to 1990 levels, for example) and the challenge of decarbonising some difficult to reach industrial sectors. Sectors of concern, in this regard, include steel making, cement manufacture, plastics, heavy transport (road and marine) and aviation.

In this book, we focus on the potential benefits of using hydrogen as an energy carrier and thus far we have considered the possible role to be played by hydrogen in meeting the challenges of future mobility and domestic heating. We have further considered that, with carbon capture, utilisation and storage, hydrogen might continue to be sourced from fossil fuels (especially natural gas) but with a limited greenhouse gas emissions impact. As regards the more difficult to decarbonise sectors, we have already commented on the benefits of hydrogen in heavy ground transport (road and rail), and we can add the potential for hydrogen cryomagnetics (Chap. 9) to make possible an all-electric fan engine for passenger aircraft, involving compact lightweight high torque superconducting electric motors.

© Springer Nature Switzerland AG 2020
W. J. Nuttall and A. T. Bakenne, *Fossil Fuel Hydrogen*,
https://doi.org/10.1007/978-3-030-30908-4_10

In this chapter, we go further and look in some detail at one of the most significant of the supposedly difficult to decarbonise sectors—steel production, and we see a significant opportunity for hydrogen-based solutions.

10.1   The Environmental Impacts of Steel Production

Seen holistically, i.e., including material transport, etc., global steel production is responsible for roughly 6% of total greenhouse gas emissions [1]. Steel is produced commercially in one of two ways: blast oxygen furnace or electric arc furnace. The electric arc process tends to be used in the production of lower quality steels and/or for steel recycling. Globally, steel recycling rates are high (83%, reaching 90% in some territories) [2]. Recycling, however, is far from sufficient to meet ever-expanding global demand. Indeed, global annual steel demand is expected to grow by 30% by 2050 [3]. There is a third route to steel production—the open-hearth process. For much of the twentieth century, the open-hearth process was the preferred means to produce high-quality steels. However, in recent decades, it has been overshadowed by the two dominant methods introduced above, and now, it is responsible for just one-sixth of steel production worldwide [4].

In the early twenty-first century, the dominant process for the manufacture of new high-quality steel is the blast oxygen furnace process. The process not only dominates in terms of steel production (approximately 65% [1]), but also in terms of global greenhouse gas (GHG) emissions—see Fig. 10.1.

The blast oxygen furnace, also known as the basic oxygen furnace, or the Linz–Donawitz-steelmaking process, utilises a water-cooled lance to inject high-pressure oxygen into a mix of molten pig iron, lime and other fluxes. The temperatures and oxygen flux are sufficient to oxidise impurities of carbon, silicon, manganese and phosphrous [5]. New blast oxygen furnace steel production relies on a feedstock of pig iron produced by a smelting furnace. Pig iron is a very brittle form of iron (with a high carbon content) produced solely as an intermediate step in the process of steel manufacture. It is the smelting (i.e. the chemical reduction) of iron oxide rich ores to metallic iron that represents a major greenhouse gas emissions challenge. Pig iron is produced in a blast furnace containing coke and fluxes. Coke is coking coal from which impurities have been driven off by coke-oven processing. The resulting coke comprises almost pure carbon. In the blast smelter, air is driven in, and this burns the coke forming carbon monoxide (CO). The fire melts the iron ore, and the carbon monoxide reduces it to metallic iron (pig iron). The smelting process requires an input of coking coal for two purposes first to provide heat and second to generate a reducing agent (CO) to convert iron oxides into metallic iron. The use of coke is an extremely severe source of harmful greenhouse gas emissions as the reducing process involves the oxidation of the carbon monoxide to the greenhouse gas carbon dioxide. One key path to lower carbon steel production would be to achieve the necessary smelting (reduction of iron ore) by means that avoid the use of coke and its associated GHG emissions.

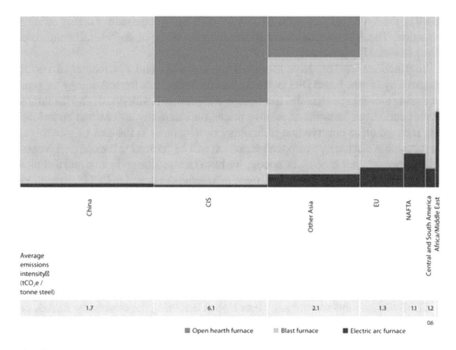

China   CIS   Other Asia   EU   NAFTA   Central and South America
                                                                                          Africa/Middle East

Average
emissions
intensity
(tCO₂e /
tonne steel)

| 1.7 | 6.1 | 2.1 | 1.3 | 1.1 | 1.2 |

0.6

■ Open hearth furnace   ▦ Blast furnace   ■ Electric arc furnace

Fig. 10.1 Global steel production GHG emissions by production method and territory, data correct at time of publication in 2011. *Source* Carbon Trust [1]

10.2   The Prospect of Low Emission Steel Production

A major theme of the Energy Transitions Commission's 2018 *Mission Possible* report concerns the ways in which the greenhouse gas emissions impact of steel production might be significantly reduced [6]. Four fundamental suggestions are presented: a move to heating by direct electrification; the use of biomass as an energy carrier and as a feedstock (e.g. biomass-coke); the use of carbon capture utilisation or storage and the use of hydrogen as an energy carrier and reducing agent. Given the scope of this book, we shall now turn our attention to the last of these four ideas.

Hasanbeigi and co-workers have provided a comprehensive review of modern and prospective steelmaking techniques [7]. As one of several future options, they assess the prospects for hydrogen as a reducing agent in iron smelting. While hydrogen-based direct reduction of iron (DRI) is currently cost prohibitive, it holds out various technical and environmental advantages. Hasanbeigi et al. point to the potential for 96% GHG emissions reduction in iron production compared to conventional carbon monoxide based blast furnace methods. The use of hydrogen would also be a blast furnace method with the oxidation of the hydrogen reducing agent yielding water in pace of carbon dioxide from coke-based processes. The smaller size of the hydrogen molecule compared to carbon monoxide facilitates penetration into the iron ore and better chemical contact. Hasanbaigi et al. report on experiments indicating the rapid speed of the hydrogen process with only seconds being required for highly efficient reduction. These considerations motivate the possibility of an "overall continuous direct steelmaking process".

Vogl and co-workers have modelled the technical and environmental performance of hydrogen-based DRI technologies concluding that the technology "appears promising assuming successful technology and process development and favourable market conditions in terms of relative prices for electricity and carbon emissions" [8]. They go on to observe that technology development in the area of electrolysis for renewable electricity hydrogen generation will be "crucial to the competitiveness of the process". In this book, of course, we have stressed the economic and technical advantages of fossil-fuel-based hydrogen production, matched to CCUS capabilities for maximal cost-effective environmental benefit.

Fischedick et al. [9] have also considered the economic and technical credentials of future steelmaking methods adopting a scenarios-based approach envisaging the prospect of hydrogen-based direct reduction (based on renewable Green Hydrogen) from roughly 2050 onwards. We speculate whether more Mature Hydrogen production technologies might be able to accelerate the decarbonisation of steel production ahead of such timescales.

The Swedish joint venture known as Hybrit involving the steelmaker SSAB, the electricity company Vattenfall and the iron ore producer LKAB has been formed to develop hydrogen-based low carbon steel production [10]. The proposition is to use Green Hydrogen produced from otherwise surplus renewable electricity. The plan is to develop a pilot plant near the northern Swedish town of Lulea close to iron ore resources. The aim is to have the plant up and running by 2020.

If such developments are successful and expecting that global concern for the damaging consequences of climate change grows in the coming years, then one can expect steelmaking and other high GHG emissions processes to come under severe policy pressure. In such scenarios, hydrogen could have a vital role to play. Currently, in Europe, there is a linkage with moves in favour of Green Hydrogen as part of a renewable electricity-based future, but in this book, we point to an alternative way ahead based largely on natural gas which could prove easier and cheaper and have similar environmental benefits. The main purpose of this chapter, however, has been to point to the fact that if the hydrogen economy grows in the years to come (for any reason), then beneficial consequences can be expected to emerge in a wide range of industrial and societal domains. A growing hydrogen economy has the potential to deliver benefits across a wide range of sectors. In this book, we stress the possibility that such changes might be delivered, not as a consequence of an expansion of renewable electricity, but rather as part of an evolution of the established fossil fuel companies' strategies assisted by the industrial gas companies and focussed on achieving greatly reduced GHG emissions primarily via CCUS. Such issues will be considered further in Chap. 11.

Acknowledgements We gratefully acknowledge the advice of Charles Forsberg, the responsibility for what follows lies with the authors alone.

References

1. The Carbon Trust, International Carbon Flows—Steel, May 2011, Ref CTC791. Available at: https://www.carbontrust.com/media/38362/ctc791-international-carbon-flows-steel.pdf.
2. Energy Transitions Commission, *Mission Possible,* Adair Turner (Chair), November 2018. Available at: http://www.energy-transitions.org/mission-possible.
3. Energy Transitions Commission, ibid. Exhibit 2.2 p. 56.
4. Encyclopaedia Britannica,Open Hearth Process. Available at: https://www.britannica.com/technology/open-hearth-process.
5. Encyclopaedia Britannica, Basic Oxygen Process. Available at: https://www.britannica.com/technology/basic-oxygen-process.
6. Energy Transitions Commission, *Mission Possible,* Adair Turner (Chair), November 2018, available at: http://www.energy-transitions.org/mission-possible.
7. Hasanbeigi, Ali, Marlene Arens, and Lynn Price. 2014. Alternative emerging ironmaking technologies for energy-efficiency and carbon dioxide emissions reduction: A technical review. *Renewable and Sustainable Energy Reviews* 33: 645–658. https://doi.org/10.1016/j.rser.2014.02.031.
8. Vogl, Valentin, Max Åhman, and Lars J. Nilsson. 2018. Assessment of hydrogen direct reduction for fossil-free steelmaking. *Journal of Cleaner Production* 203: 736–745. https://doi.org/10.1016/j.jclepro.2018.08.279.
9. Fischedick, Manfred, Joachim Marzinkowski, Petra Winzer, and Max Weigel. 2014. Techno-economic evaluation of innovative steel production technologies. *Journal of Cleaner Production* 84: 563–580. https://doi.org/10.1016/j.jclepro.2014.05.063.
10. Frederic, Simon. 2018. *Swedish steel boss: 'Our pilot plant will only emit water vapour'.* EURACTIVE.com, May 11, 2018. Available at: https://www.euractiv.com/section/energy/interview/hybrit-ceo-our-pilot-steel-plant-will-only-emit-water-vapour/.

Chapter 11
The End of Oil—Hydrogen, Syn-Fuels and the International Oil Companies

> Thirty years from now there will be a huge amount of oil - and no buyers. Oil will be left in the ground. The Stone Age came to an end, not because we had a lack of stones, and the oil age will come to an end not because we have a lack of oil.
>
> Sheikh Yamani former Saudi Oil Minister speaking in June 2000 [1]

Sheikh Yamani's insight is proving to be remarkably prescient. The end of oil is now in sight, and the driver of that reality is not being driven by any real shortage of oil.

Oil has been special in energy economics because arguably it is the only energy resource with a credible status as a true market commodity. Of course, there has been a long history of attempts at achieving market power in oil. It has to be conceded that for the first three quarters of the twentieth century, there was no real market in oil, rather there was a western oligopoly of producers known colloquially as the "Seven Sisters"—a classical allusion first coined in the 1950s by Enrico Mattei, the former boss of the then Italian state oil company ENI. The allusion is to the seven Pleiades, daughters of the Greek god Atlas. The Seven Sisters oil companies were [2]

From Europe:

- Anglo-Iranian Oil Company, later British Petroleum
- Royal Dutch Shell

And from the USA:

- Gulf Oil
- Texaco

Both companies were later merged into Chevron.

The three remaining sisters were also from the USA, and they were descendants of the Standard Oil empire established by John D. Rockefeller

- Standard Oil Company of California
- Standard Oil Company of New Jersey
- Standard Oil Company of New York

© Springer Nature Switzerland AG 2020
W. J. Nuttall and A. T. Bakenne, *Fossil Fuel Hydrogen*,
https://doi.org/10.1007/978-3-030-30908-4_11

The California company would later find its way into Chevron while the New York and New Jersey enterprises are today part of ExxonMobil.

In the 1970s, OPEC attempted to achieve dominance on the oil supply side and the International Energy Agency was created in response to represent the concerns of the demand side. The high point of such efforts was, however, now more than forty years ago, and in recent years, no-one player, supplier or buyer, has been able to "make the price" of oil, be it Brent Crude or West Texas Intermediate; rather for the most part everyone in the market has been a "price taker".

Oil is an energy-dense liquid with low transportation costs and hence it was perfectly suited, in the late twentieth century, to be traded in a global market. One cannot point to any other form of commercial energy that has been allocated via a global market to such an extent. Of course, natural gas and electricity markets have both been attempted, and some might point to examples of market price formation and resource distribution, but generally, however, such markets have been undermined by policy interventions and physical constraints, such as, for example, those arising from market-inappropriate legacy gas pipeline infrastructures.

While electricity transformed civilisation during the twentieth century, it was oil that transformed mobility: from the Model-T Ford to the jumbo jet, petroleum allowed for the affordable movement of goods and people. The importance of oil was enormous, but while still substantial, it is now waning and the reasons are numerous. Fundamentally, the change is coming not from the classic concerns around "peak oil" as arising from a peak in supply and the need to adjust to looming scarcity. Rather it seems increasingly likely that the end of oil will be driven by a decline in demand. The need to reduce greenhouse gas emissions and the consequent shifts in technology, for example, to electric vehicles, are important considerations motivating such an impression. The end of oil will be associated with abundance and low prices, not scarcity and high prices. A consideration not to be neglected is who will exert control over, and have ownership of, the oil that remains. We suggest the latter issue is at least as influential when considering the end of the age of oil as it was previously.

The Seven Sisters evolved, and it is more usual today to speak of the international oil companies or IOCs, namely Shell, BP, ExxonMobil, etc. They are powerful players in western capitalism. Indeed, the chances are that your pension plans will have some significant linkage to their long-term success. In recent decades, they have gradually been moving away from oil to become natural gas companies. Their broadened resource base has allowed them to point to a lower carbon future as carbon-intensive oil is replaced by less environmentally damaging natural gas. The carbon benefit, however, is more a fortuitous consequence of strategic choices largely made for other reasons, many of which relate to declining access to upstream oil.

The oil majors are familiar to us all, and at one time, they dominated global oil supply. In 1972, Anthony Sampson estimated that together the Seven Sisters controlled 70% of global oil production [1]. Much of this activity related to joint undertakings with national oil companies, but prior to the 1973, first oil shock control most definitely lay with the western firms. Today, however, things are very different. Based on a 2010 assessment, today's international oil companies control less than 10% of proven oil reserves and less than 25% of global oil production [3]. In contrast, the vast majority of global oil is now firmly controlled by national oil companies, such as Saudi Aramco and Russia's Rosneft. As Sheikh Yamani's prescient insight implies, we should not expect the end of oil to be associated with high and rising prices. After all, such scenarios hardly represent an incentive for an incumbent to exit the business. The end for oil is far more likely to be associated with a period of sustained low oil prices than a period of high, and rising, prices. This is a key aspect of the peak in demand coming before the peak in supply. In 2017, oil supply was relatively abundant following the return of Iran to the market following the global lifting of nuclear sanctions in January 2016, although this particular new dawn appears short lived in the era of President Trump with his foreign policy arguably aligned with the priorities of Saudi Arabia and Russia and against the interests of Iran. Despite these actions, one can still see the prospect of a future based on notions of oil abundance rather than oil shortage. The moves in the USA by the "fracking" industry away from gas to more lucrative tight oil have further added to the global supply-demand imbalance. Oil is abundant while the world's interest in it is nevertheless starting to move away. One important aspect of that new aversion has been the diesel emissions scandal discussed in Chap. 1.

These trends pose a serious threat to the oil majors as they compete with the national oil companies (NOCs). Generally, all the world's "easy oil" is in the hands of the NOCs, and oil still gushes out of the ground for Saudi Aramco so easily that for the Saudi's the breakeven oil price is around $15 per barrel (Fig. 11.1). Meanwhile, the oil majors scramble from one difficult place to the next attempting to preserve their reserve replacement ratio and to get their hands on some good "equity oil". Unfortunately, however, this oil tends to be from the frozen north, or the deep sea, and it is never obtained cheaply. In a world where the oil majors can only access oil at a minimum cost of $45 a barrel, then if crude oil is trading at $45, then on such days they are making no money at all. In contrast, their NOC competitors are typically still making up to $30 a barrel. Such a situation appears unsustainable and will further erode the relative position of the IOCs with respect to the NOCs. Of course, many countries with national oil companies have deep budget deficits and would much prefer the oil price to be higher. But in a world with no market power, they are unable by themselves to cause prices to rise. As a consequence, the indebted producer nations have only one short-term rational strategy: to pump as much oil as possible so as to maximise revenue on the back of volume despite the inevitable further downward pressure on price.

We argued earlier that the international oil companies are increasingly international natural gas companies. If the end of oil is indeed in sight, then further strategic shifts for these companies can be envisaged.

In his 2017 book *Burn Out*, Dieter Helm presents a persuasive assessment of the big picture for energy, both in respect of fossil fuels and electricity [4]. To our impression, Helm's core argument runs as follows:

• The prospect of enduring low oil prices brings with it the probability that the energy game will change profoundly to the detriment of energy companies sticking with long-established strategies.

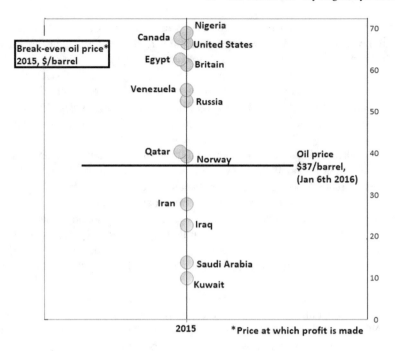

Fig. 11.1 Breakeven oil prices for various producing countries in 2015, adapted from information published in The Economist, 8 January 2016

- Renewables, and most especially first generation renewables, have for those countries leading the way proven to be more problematic than advantageous.
- The key short-term problem to be addressed is to end the unabated combustion
  of coal. It is unfortunate that German enthusiasm for first-generation renewables
  (and we might add German nuclear aversion) has given a boost to coal combustion
  in Europe.
- Fundamentally, the future of the electricity system must be based around the economics of an increasingly zero marginal cost business. The future electricity system
  will look more like broadband Internet with payments for access and not payments
  for usage. In such a future, it will be absurd to argue that we should worry about
  "wasting electricity", just as few today worry about wasting broadband data.
- Electrical energy will be free, but access to it will not. Such a shift will have consequences all the way up the energy value chain, and it will alter business models and
  strategies profoundly. Helm posits that the era of large energy companies absorb-
  ing risks internally is probably gone. IT and data transparency will allow a future
  built by smaller specialised players doing business with each other collectively to
  serve end-customer needs.

We tend agree with Helm in his views on such matters, but we are intrigued by
the possibility that he might be wrong when, to our understanding, he further posits
that:

- Gas will be a transition fuel and that it has no significant long-term future in a decarbonised world.
- That carbon capture and storage is unlikely to make a material difference to our energy futures.
- The electrification is the way to see the global energy future—to the detriment of both established electricity companies and fossil fuel companies.

Rather we see a possibility, in mobility at least, that the international oil companies in their new form as international natural gas companies, will evolve so as to lead a hydrogen economy for next-generation fuel cell electric vehicles. In many ways, this future will appear more natural and less disruptive to the user experience than the vision of electrification advanced by Helm.

In essence, we posit that the former Seven Sisters companies should complete a serious of evolutionary steps that have seen them already transition from being major oil companies to being major natural gas companies, and in time, we hope they will further transition to being major low-carbon fluid syn-fuel companies with a special emphasis on hydrogen and possibly ammonia.

We put forward our hypothesis that the international oil companies and the industrial gas companies might be able to engineer a low carbon future in an evolutionary way based largely on natural gas processing. In doing so, we draw upon a sense that much of the thought leadership around energy policy in Germany and the European Union has been built upon a series of axioms that are perceived by the adherents to be virtuous, but which are, at their core, are actually somewhat independent of energy policy, namely

- The future should be local and in the hands of citizens rather than large-scale, national and in the hands of experts, corporations and national politicians.
- That nuclear energy is inherently evil and that this is revealed by the linkages to radiation-induced cancer and nuclear weapons—noting the mass casualties at Hiroshima and Nagasaki and the real risks of nuclear weapons proliferation.
- Energy is expensive and should be utilised with minimal waste.

However, economic and engineering thinking rather points in different directions, as this book has discussed.

European green-thinking these-days involves much enthusiasm for renewables (and for many the smaller and more local the better). Arguably this enthusiasm links to a very long-standing aversion to energy futures built by large corporations. The origins of renewables advocacy were in the early 1970s spurred by a convergence of anti-corporate ways of living with a desire to insulate society from primary fuel energy insecurity, a concern that grew greatly during the first oil crisis. Only later did climate change become a motivating factor. The desire for anti-corporate self-sufficiency can still hold much attraction. The countervailing notion that our low-carbon future might be built by large engineering-based businesses can have have echoes for some of President Eisenhower famous warning of a "military industrial complex" [5]. We suggest that much of the popular enthusiasm for renewables resides in a philosophical positioning rather than from technical analysis.

In *Burn Out*, Helm explains well the logical fallacies in some of the green rhetoric of recent years stressing the importance of renewables (which have historically proven to be expensive and which have frankly advanced, rather than diminished, the cause of coal in Europe) and in favour of efficiency (which would be no virtue at all in an electricity, and increasingly energy, future based on zero marginal cost supply).

To be clear, as authors of this book, we are guided by two dominant concerns. First, we have a genuine concern for global climate stability and insist that GHG emissions are reduced dramatically and quickly. Second, we have a sense that energy is a technically difficult area of activity best left, at some level at least, to the experts. Of course, there is a need for societal understanding and acceptance, but this should not be achieved at the expense of over-simplification of the issues.

In this book, we point to an alternative way ahead for the international oil (now natural gas) companies and the industrial gas companies to lead the world into low-carbon mobility. In our scenario, the wheels of future vehicles will be turned by electricity, but that electricity will be made on the vehicle from clean hydrogen supplied from natural gas, and today's electricity industry will have very little to do with it. This will be a mobility future built by established oil companies extending the dance marathon they have enjoyed with the car companies for one hundred years. This will not be a story of technological revolution, it will more be one of technological evolution. By such means, we can achieve material and rapid emissions reductions and generate few new secondary problems along the way.

Helm and others posit the beneficial role that battery electric vehicles could play in balancing an ever more volatile electricity system (packed full of intermittent renewables), but why would those in the power business, want to associate themselves with tricky problems that currently lie outside the domain of electricity and which look like they are only getting worse and ever more political [2]. Furthermore, the challenges facing electricity system renewal and decarbonisation are already substantial. Rather let us recognise that hydrogen fuelling potentially represents a much simpler story of stored energy, very similar to the biggest energy storage activity on the planet today—the global petroleum supply industry with its storage at innumerable filling stations together with the half-full petrol tanks of all the world's car, buses and trucks. Just as the electricity sector faces no obligation to fix problems in mobility, then arguably the mobility sector faces no special or moral obligation to help fix the electricity sector's emerging difficulties of supply-demand balance. Rather the mobility sector could (and indeed arguably should) carry on efficiently meeting its own consumer's needs but in significantly more environmentally responsible ways. This might be done via a shift from fossil fuels on the vehicle to clean hydrogen. Clean hydrogen has no tailpipe emissions and very low whole lifecycle emissions if sourced from natural gas and if the technology of production is linked to carbon capture utilisation and storage (see Chap. 5).

In this book, we have made the case that the future hydrogen economy should be based primarily on the clean processing of fossil fuels, rather than on renewables. At its heart, the argument for that goes as follows:

- Low-emission abated natural gas sourced hydrogen appears currently to be significantly cheaper than any renewables-based alternative (Chap. 4)
- The environmental impacts (land-use, emissions and air quality) of the natural gas extraction industry may have the potential to be similar to, or perhaps even better than, the impacts of solar power, wind power or biomass production and transport.
- The technical capabilities and financial strength of the international oil (gas) companies render credible the looming technological evolution, noting that technological evolution is usually easier and cheaper than revolution.

Finally, if the current (at time of writing) era of relatively low oil prices persists, then international oil (gas) companies must shift from being low technology resource companies to being more technologically sophisticated and ambitious. They should not do this simply because it is virtuous, which it is, but also they should do it with a view to their own survival. They should "go green" and make the associated public affairs volte-face. In the future, they should lobby aggressively to block, or render uncompetitive, old-fashioned high-emission petroleum fuels from the market. In this way, they might inhibit the lower-tech national oil companies as they seek to enter into key global retail markets (essentially the OECD countries) on the back of ever higher control of upstream oil assets, When faced with losing that game, the international oil companies should change the game, and by good fortune the game that gives them a source of strength is the move (via technology and innovation) to low carbon fuels—such as hydrogen.

Selling hydrogen and other low-carbon fluids is a much more attractive and natural business proposition that seeking to join the electricity sector with its very different norms and culture.

Acknowledgements With thanks to Nick Butler for his generous insights, but the opinions expressed are the responsibility of the authors alone.

References

1. Fagan, M. 2000. Sheikh Yamani predicts price crash as age of oil ends. *The Telegraph*. It is interesting to note that this aphorism appears to predate Sheikh Yamani's use of it. See: https://quoteinvestigator.com/2018/01/07/stone-age/.
2. Johnson, H.L. 1976. ANTHONY SAMPSON. The seven sisters: The great oil companies and the world they shaped. *The Annals of the American Academy of Political and Social Science* 425 (1): 156.
3. Tordo, S., B.S. Tracy, and N. Arfaa. 2011. *National oil companies and value creation*. National Oil Companies and Value Creation, W.B.W. Paper.
4. Yearley, S. 2017. *Dieter Helm: Burn out: The endgame for fossil fuels*, vol. 9, 283. Yale University Press. ISBN 978-0-3002-2562-4.
5. Eisenhower, D.D. 1961. Farewell address to the nation. In *Video and commentary on the origins of the famous speech via the US National Archives*.

Chapter 12
Conclusions

In this book, we have sought to examine the economic, technical and environmental issues surrounding the proposition that hydrogen from fossil fuels might play a transformative role in decarbonising the global energy system. We have considered issues around the production of hydrogen from fossil fuels and stressed the importance of moves towards lower environmental impact, especially concerning greenhouse gas emissions. We have argued that hydrogen could play a substantial role in the future of low-carbon mobility and also in domestic heating in countries, such as the UK, currently reliant on natural gas supply to end-user consumers.

Throughout the book, there has been an alternative future vision to be considered—the possible large-scale expansion of electricity into heating and mobility based on renewable (and possibly nuclear) power generation coupled with energy storage. One candidate for such energy storage is the use of hydrogen produced using otherwise surplus renewable electricity from water by electrolysis.

12.1 The Renewable Green Hydrogen Alternative

Around the world, we are seeing growing enthusiasm for renewable energy-based policies coupled with falling prices for the manufacture, and even installation, or renewable energy technologies. Consequently, the world is seeing a rapidly rising proportion of renewable energy in the electricity system. That said, however, global electricity production is still dominated by fossil fuel combustion, based largely on coal, but increasingly on natural gas.

As the renewables contribution to global electricity capacity rises, there are growing chances that there will be times when in a given region electricity supply exceeds conventional demand. Despite efforts to shift such imbalances in time (via storage) and in space (via grid interconnection), there is still the risk that the future electricity system will exhibit very low (or even negative) spot market prices and risk the significant "waste" of unwanted electricity. Although as discussed in Chap. 11, in future,

© Springer Nature Switzerland AG 2020
W. J. Nuttall and A. T. Bakenne, *Fossil Fuel Hydrogen*,
https://doi.org/10.1007/978-3-030-30908-4_12

such unwanted electricity may no longer be seen as a waste. Indeed, if renewable electricity is to make the time-averaged contribution that some are calling for (even up to 100% of all electricity), then it is inevitable that there will be substantial periods of time where renewable power generation (produced at effectively zero marginal cost) will exceed demand requirements. Those positing a strong role for Green Hydrogen suggests that moves to large-scale storage can reduce the risk that electricity will be wasted. For example, via electrolysis-generated Green Hydrogen-based energy storage, some beneficial use can be made of the otherwise surplus, and in marginal cost terms, "free" renewable electricity. Of course, in full cost terms, the electricity is far from free, but that is another matter. If such policy trends continue, one can assume that the prior construction of renewable power generating assets will have been assisted by subsidy and a socialisation of the costs across the electricity sector. That said such costs are falling fast and one can imagine scenarios in which subsidies might not be required. Looking ahead, it is hard to envisage the nature of the future electricity business, especially if it shifts from the current reality in which supply and demand must be balanced in real-time and while large-scale storage still remains an idea for the future. While the development of large-scale storage will require significant policy support, one can imagine a future in which renewable generation assets and associated storage might be funded in ways that become the new normal and the concept of subsidy will not really apply. Looking ahead at possible futures, we are conscious of Dieter Helm's ideas concerning a future in which electricity storage does not increase to meet the system balancing needs associated with large-scale renewables (see Chap. 11). He makes the suggestion that if significant renewable expansion proceeds without large-scale storage, then electricity bills might be associated with an electricity connection rather than the energy supplied. This will be because electricity waste will be so ubiquitous that it will not be regarded as waste at all [1]. A key aspect of there being no waste in such a scenario is that the electricity in question will have been generated at zero marginal cost. If, in addition, it has zero value, then it is not surprising that it will have zero-price. In such circumstances, it is hard to argue, economically at least, that there is any waste.

The prospect that future surplus renewable electricity might be stored via the electrolysis of water to hydrogen has prompted much interest in a range of related propositions collectively known as "Power-to-Gas". For example, Power-to-Gas Green Hydrogen might be stored for later use in a stationary fuel cell, i.e., as a pure electricity storage device, or it might be blended with natural gas and sent into the public natural gas distribution network. Power-to-Gas thinking can also be extended to a range of other gaseous fuels, up to and including the production of synthetic methane (equivalent to natural gas). However, one of us (WJN) has long regarded the proposition of converting renewable electricity to methane as somewhat bizarre, likening it to some form of "inverse alchemy". The idea that one might devote significant chemical engineering effort to convert electricity into methane is frankly odd, when compared to the more orthodox activity of converting methane to electricity. The reference to "inverse alchemy" alludes to the idea that one is converting valuable gold (electricity) into cheap and abundant lead (natural gas). It may be an intellec-tually interesting challenge for subsidised chemical engineering researchers, but all concerned should pause to ask: *is it really a good idea?* Surely a better, and seem-ingly easier, avenue for researchers would be to work to avoid the GHG emissions currently associated with methane combustion.

The electricity system appears to be in transition from an economic framework based upon fuel costs and significant marginal costs of generation to one dominated by capital costs and near-zero marginal costs, but many things remain uncertain. How markets will be redefined in the face of that reality remains to be seen. It has long been posited that energy costs will fluctuate in respond to immediate supply-demand mismatch, but this is not inevitable. Twenty years ago many people might have expected that the electricity transition away from fossil fuels might have been associated with very high fuel costs as resources became harder to access. Such narratives had much to do with the concept of "Peak Oil" [2]. In Chap. 11, we considered the prospect that in the future units of electrical energy could actually be free. This harks back to the aphorism "too cheap to meter", but not for the reasons for which that phrase was originally invoked [3]

Let us, however, return to considering more proximate futures: short term-prices may occasionally be low, but on the average, the new electricity system will be far from cheap. We suggest that the decisions concerning the future of our electricity system, and how we ensure that the system is decarbonised, would be better made with a whole-system strategic overview so as to be able to minimise total costs and to ensure maximum system reliability. However, we concede that such remarks might be seen perhaps as little more than a call for a return to the days of technocratic electricity monopolies and national-champion-advocated industrial policies. We concede the waste and poor decision-making of those former days, and we can see the beneficial technological innovations (especially relating to renewables) of recent years. That innovation has not just been a consequence of large subsidies it has also been a consequence of bottom-up firm-led initiatives advanced by "learning by doing" [4]

Perhaps the Power-to-Gas production of methane from electricity will be just another step in a sequence of beneficial innovations, and in this way, we will see maximal environmental benefits from low-carbon renewables. Over the long-term, costs may fall (as they have for renewable power generation—first solar and then wind) and the whole system economics of such ideas will in future be far preferable to today's realities. Indeed, we must point to existing progress towards the Power-to-Gas concept; for example, the European BioCat consortium led from Denmark is developing a 1 MW power-to-gas demonstration rig [5]. The aim is to use water electrolysis to make hydrogen. The hydrogen is then converted to methane in a bioreactor in which the process is driven by the actions of microorganisms. The resulting methane gas is then distributed in the Copenhagen region via the standard gas distribution grid. Of course, as discussed elsewhere in this book, such schemes today face an uphill challenge in overcoming the cost advantages of fossil fuel natural gas. If natural gas prices can be expected to rise, then the fortunes of renewable power to gas might improve, but it is far from clear that cost increases do lie ahead for natural gas. In this book, we tend to the view that the long-term prospects for natural gas are increased abundance and lower prices. Even if US-pioneered shale gas innovations, such as "fracking" remain barred in several EU member states, global gas market dynamics will ensure that European countries see falling prices. Such globalisation of gas prices will be favoured by the expansion of liquefied natural gas to complement existing pipeline distribution networks. The moves by the USA to become a major liquefied natural gas (LNG) exporter, unimagined 20 years ago, will further drive this dynamic.

The Power-to-Gas proposition is associated with an electricity system dominated by renewables with such renewable generation capacity installed that significant surplus power can be expected. While such power might be sold at a low price, overall renewables capacity must be paid for by someone. Today, these costs are usually socialised across electricity consumers. In the UK, this is seen for both renewables and nuclear power via "Contracts for Difference". It remains to be seen how consumers and voters will respond to such socialised costs as they see the policy component of their electricity bills rise, especially if the electricity sector takes on wider decarbonisation challenges.

The consultancy firm Ecofys observed in February 2018, in a report for Gas for Climate a consortium of seven European gas transmission and distribution companies, that by 2050 EU "green gas" (comprising biogas and hydrogen from renewables) could total 120 billion cubic metres (bcm) [6]. This quantity immediately appears impressive, and indeed the report further argues that this could save Europe €140 bn each year. These impressive sounding figures should, however, be put in some context. The European Commission reports that, for 2017 as a whole, EU total consumption was 491 bcm. This means that the ambitious goals reported by Ecofys for 2050 represent less than 25% of today's gas demand. In a context in which greenhouse gas emissions must be reduced by 80% overall, it seems that even the optimistic green gas scenarios will be insufficient to underpin a decisive shift in emissions. Indeed, it further clarifies the proposition that the key issue will be whether the low emissions future can be achieved by a shift away from our usage of 491 bcm of natural gas to a future based on far lower gas volumes or rather, as posited in this book, will natural gas continue to have a strong role to play shifting to CCUS and hence contributing to an ultra-low carbon future? Specifically, can a natural gas-based energy system be modified to supply hydrogen for end-user consumer use? Finally we must remember that the future of the global climate will not be determined directly the actions of the UK, but rather by the choices to be made by large and expanding economies, such as India. It is vital that socially and economically viable low-carbon pathways are found that fit the needs of people in such parts of the world. Effective, efficient and relatively low-cost solutions must not be overlooked.

12.2 Water Electrolysis for Green Hydrogen

Notwithstanding the high levels of socialised cost and the somewhat unattractive economics overall, it is nevertheless the case that for investors a business case can be made for renewable energy storage via Green Hydrogen. It is important to concede that part of this business case arises from the success that has been achieved in driving down the cost of various renewable generation technologies (especially solar photovoltaic panels) in recent years. As a consequence of the various changing factors in cost allocation, subsidy and capital equipment cost, electrolysis for hydrogen production is rapidly becoming more interesting to technologists and investors. Electrolysis may be performed using a range of technical means. High-temperature steam electrolysis would appear to be the most attractive at present. George Tsekouras and John Irvine have considered these issues as part of their work for the DOSH2 consortium [7], and their insights are complemented by input from Andrew Cruden in the same consortium. Together with NEL Hydrogen, Rand Technical Services in South Africa has been working in recent years to produce a small hydrogen electrolyser: the NEL P-60 [8]. Their goal is to meet the challenge of end-user affordability.

British-based hydrogen company ITM Power has achieved much success in advancing the production of Green Hydrogen from renewable sources. It has a major presence in the emerging Power-to-Gas industry in Europe, and it has the innovative HGas product. The HGas offering fits within an ISO shipping container, and it comprises a self-pressurising PEM electrolyser. The system is optimised for rapid response so as to favour the use of short-term surplus renewable electricity [9] (c.f. Chap. 8).

Beyond electrolysis, there are other more advanced water separation technologies with the potential to be combined with renewable energy. One example is the Bio-Cat biological approach in Denmark mentioned earlier [5]. Another approach is the direct high-temperature catalytic thermochemical splitting of water; one example of which is the very high-temperature (750 °C and above) Sulphur-Iodine (SI) cycle, as pioneered at laboratory scale by General Atomics in San Diego [10]. In contrast to electrolysis, the SI process splits water without the need for substantial amounts of electricity. Fundamentally, the process can be driven by any high-temperature heat source. Concentrated solar energy, as supplied by heliostat systems, or similar, in high solar insolation areas (e.g. Spain or North Africa) would be one way in which to generate the very high temperatures required. Other thermo-chemical approaches, such as the Calcium-Bromine cycle, have the benefit of being possible at lower temperatures, but some electrical energy is required in these cases. It must be pointed out that the SI process involves some very challenging chemical engineering including the use of concentrated sulphuric acid and hydrogen iodide at very high temperatures.

In passing, we observe that the production of hydrogen by water electrolysis in principle yields an important co-product: oxygen. At present, however, the typical proposition is for this oxygen to be vented and lost. It would feel less wasteful if something useful could be done with the potentially valuable substance. We considered such ideas in Chap. 5 (e.g. oxygen for oxy-fuel fossil fuel combustion with CCS). One possibility, applicable even in areas without significant chemical process industries, would be to use it in metropolitan waste water treatment.

In this book, we suggest that an alternative paradigm based on the emergence of a similarly environmentally sound hydrogen future to the electrification vision but based on fossil fuels with CCUS. We posit that such a future might be progressed without substantial subsidies. Indeed, an appropriate carbon price alone might be sufficient. We suggest that the necessary carbon price lies within the reach of economically and politically acceptable limits (see Chap. 4). Given observed experience with carbon prices around the world, it would be important to ensure the long-term credibility and materiality of such a policy proposition. We further note that a fossil-fuel-based hydrogen future has the potential to be led by highly experienced companies and expert entities, such as the international oil companies and the industrial gas companies. These companies have a wealth of relevant experience to draw upon.

12.3   Electricity Transition or Energy Transition?

Thus, far in this chapter, we have tended to consider the role of hydrogen as an electricity storage capability in an electricity system replacing fossil-fuelled generation with intermittent renewable sources. As discussed earlier in this book, however, that is merely a small part of the possible future for electricity. There will be increasing demands on the UK electricity system arising from the electrification of mobility, as discussed in Chap. 2 and even greater demands if electricity is to play a dominant role in domestic heating. As things stand today, most UK homes are heated using natural gas (see Chap. 7).

It is not the purpose of this book to imply that renewables-based green futures are impossible. Indeed, if the issue at stake is simply a limited evolution of the electricity sector serving current goals that possibility seems almost within reach today. Rather our intention is to pause and reflect on the wider decarbonisation challenge including mobility and heat and extending into hard to decarbonise sectors (as discussed in Chap. 10). With these ideas in mind, it seems increasingly clear that renewable electricity alone will be incapable of meeting our needs in an affordable and efficient manner. These realities combine to reinforce our sense that other energy sources might be needed. It might be that a nuclear renaissance is required, and one of us (WJN) has commented on such matters elsewhere. But even together, it seems that renewables and nuclear will be far from sufficient. It is the further sense that more is required that has led us to consider the potential for hydrogen from fossil fuels, developed in environmentally responsible ways. We draw upon the lessons of the late David MacKay in tending to the view that frankly all good ideas should be deployed concurrently. This matches to ideas associated with the stabilisation wedges concept advanced by Robert Socolow and Stephen Pacala.

12.4   Time to Disinvest?

In recent years, there have been growing calls for institutional investors to break their links to fossil fuel companies on the basis that such companies have been responsible for much environmental harm and that they stand in the way of progress towards an efficient and low carbon energy system based more strongly on renewable sources.

Generally, the argument goes that fossil fuel companies should be stigmatised and from that action good consequences would flow. For example, it is suggested that investment capital should be diverted towards clean and renewable energy sources. But what about the future of the fossil fuel companies themselves? In that regard, we see two possible goals behind such calls for disinvestment and stigmatisation. We have some sympathy for the first possible goal—it is essentially that the campaign will act to push fossil fuel companies towards more sustainable thinking. i.e., to encourage them to evolve their strategy and tactics towards more environmentally responsible behaviours. The second logic is more aggressive, however, and it is a thesis that we do not support. The second logic is that the stigmatisation should be part of wider efforts to make life ever-more difficult for the fossil fuel companies until they are forced out of existence.

Dieter Helm has rightly identified a set of problems underpinning the logic of disinvestment. The first is that while it can be argued that a disinvestment campaign (against Barclays Bank for example) played a beneficial role in ending apartheid in South Africa, the business models and business dynamics of banking and fuel supply are simply too different for the parallel to be valid. Second, if one puts the international oil companies out of business, who will fill the void?—Well, it is likely to be that the national oil companies will step in, and one can see no reason why those taking advantage of an emerging gap in fossil fuel supply will be more ethical than the displaced international oil companies. Generally, we tend to the opinion that disinvestment presents more risks than benefits and it further threatens expert institutions just at a time when their technical skills are most needed (in reducing the environmental impacts of fuels for energy services), such issues relate to ideas discussed in Chap. 11.

12.5 All Hands to the Pumps!

We conclude that Mature Hydrogen could represent an enduring and potentially attractive alternative to renewables-based propositions. As outlined in the preceding chapters, we suggest a way ahead based on hydrogen from fossil fuels, especially the newly abundant natural gas arising from innovations in fracking and liquefied natural gas shipments. Such Mature Hydrogen provides substantial opportunities for significant, but incremental innovation towards a low carbon future for mobility and domesticheating if combined with CCUS.

Such developments will support beneficial innovations in vehicle technology (such as the drive to electric vehicles) and to hydrogen transmission and distribution and retail hydrogen supply.

In those cases where bulk hydrogen supply is not easily feasible via pipeline infrastructure, but where natural gas is nevertheless available, then Small Hydrogen production techniques, as discussed in Chap. 8 might be helpful. We concede that for locations far from any gas pipeline infrastructure (hydrogen or natural gas) options are more limited, but they nevertheless include truck-based shipments of hydrogen gas in high-pressure gas trailer tubes or more interestingly as a cryogenic liquid (as discussed in Chap. 9). Of course, it is in places far from gas infrastructures where Green Hydrogen electrolysis based upon local renewable electricity generation can surely be expected to be the preferred approach.

As we look across this complex landscape of options, we are minded to stress that any perceptions that the Mature Hydrogen sector is "low-tech" or inevitably greenhouse gas emissions intensive are too premature and simplistic. We suggest that in some contexts there is good reason to believe that greater progress towards a low carbon future can be made from the evolution of the Mature Hydrogen sector than from an aggressive policy push for renewables-based Green Hydrogen. We suggest that ideas become even more important when seen at a global scale. If technology-neutral incentives are applied and costs are allocated closest to those making investment choices, then we expect that the evolution of Mature Hydrogen approaches may be favoured, at least in the short-term and perhaps for the foreseeable future. The most transparent and direct way to do this would be to a global carbon tax escalating with time. During the process of writing this book we have seen global interest in the potential for hydrogen grow very significantly. As the book was going to press one particularly impressive report appeared from the International Energy Agency [11]. We commend the IEA report to those interested in reading more and we concur with its authors what they conclude that "hydrogen today appears to have a tailwind".

Our closing observation is that Mature Hydrogen represents a set of experiences and ideas found largely in the domain of industry and as such it has until very recently received relatively little academic or public-policy attention. The industrial gases industry is already very close to the world of the international oil companies as it acts to support their refining and related needs. As such, there is ample opportunity for industry-led innovation around new vehicle fuel blends, i.e., lower-carbon blended vehicle fuels that are liquid at room temperature and pressure. Such a step would build directly upon existing biofuel experience such as 10% ethanol/gasoline blends. Incorporating new synthetic fuels to these blends might allow for progressively lower environmental impacts and are likely to require more hydrogen and other molecular building blocks as supplied by the industrial gas community. The primary upstream energy resource would be natural gas but the environmental impacts could be substantially lower than might naively be assumed when initially noting that the pathway proposed is fossil-fuel-based. In these closing remarks, we posit that academic and scholarly communities should do more to assess and, if appropriate, develop the responsible use of fossil-fuel-based hydrogen as a low carbon contribution to our energy future. The climate change challenge is too pressing and too important for any potentially useful approaches to be ignored.

To be clear, we are not advocating development of low-carbon Mature Hydrogen at the expense of progress relating to Green Hydrogen or to the hydrogen alternative—greater electrification. Rather we suggest that all these ideas should be advanced with full vigour. Where a competition for resources, of any type, arises we suggest that it is natural for one approach to be favoured, but that this selection should be made without prejudice or axiomatic constraints, but rather should be dedicated to meeting policy needs. We argue that highest amongst those needs right now should be a focussed on rapid and material GHG emissions reductions.

We take comfort that scientist and engineers have much to contribute to the energy challenges of our age, and despite the severe nature of the challenges ahead, we remain broadly optimistic that benign future can be found. In meeting this challenge, everyone has a role to play, but the energy engineering community has perhaps the biggest role of all.

References

1. Helm, D. 2017. *Burn out: The endgame for fossil fuels*. Yale University Press. ISBN 978-0-3002-2562-4.
2. Deffeyes, K.S. 2005. *Beyond oil: The view from Hubbert's Peak*, 204. New York: Hill and Wang.
3. Strauss, L.L. 1954. *Speech to National Association of Science Writers*, January 24, 2019. Available from: https://www.nrc.gov/docs/ML1613/ML16131A120.pdf.
4. Jamasb, T., W.J. Nuttall, and M. Pollitt. 2008. The case for a new energy research, development and promotion policy for the UK. *Energy Policy* 36 (12): 4610–4614.
5. Excess wind power is turned into green gas in Avedore. BioCat Project 2014. Available from: http://biocat-project.com/news/excess-wind-power-is-turned-into-green-gas-in-avedore/.
6. Melle, T.V., et al. 2018. *Gas for climate—How gas can help to achieve the Paris Agreement target in an affordable way*, January 22, 2019. Available from: https://gasforclimate2050.eu/files/files/Ecofys_Gas_for_Climate_Report_Study_March18.pdf.
7. DOSH-2 Project, May 27, 2014. Available from: http://www.een-northeast.co.uk/supergen-dosh2/.
8. South Africa Hydrogen Challenge Overcome, Gasworld Magazine, issue 99 (August 2013).
9. HGas Product. ITM Power, January 24, 2014. Available from: http://www.itm-power.com/product/hgas.
10. Russ, B. 2009. *Sulfur iodine process summary for the hydrogen technology down-selection*. Idaho National Laboratory.
11. International Energy Agency, The Future of Hydrogen—seizing today's opportunities, June 2019. Available at: https://www.iea.org/hydrogen2019/.

Index

© Springer Nature Switzerland AG 2020
W. J. Nuttall and A. T. Bakenne, *Fossil Fuel Hydrogen*,
https://doi.org/10.1007/978-3-030-30908-4

Printed in the United States
By Bookmasters